Birds Withou...

Investigating Populations, Habitats, and Conservation of Birds in the U.S. and Abroad

Nancy M. Trautmann – Cornell Lab of Ornithology

James G. MaKinster – Hobart and William Smith Colleges

Crossing Boundaries Teacher Leaders

Carol Burch – Hannibal High School

James Hefti – Pulaski Junior Senior High School

Lisa Lowenberg – Chittenango High School

Lisa Mason – Hannibal High School

Roberta Palmiotto – Union Springs Central Schools

Michelle Watkins – Beaver River Central High School

Crossing Boundaries Staff

Michael Batek – Hobart and William Smith Colleges
Bradley Muise – Hobart and William Smith Colleges
Courtney R. Wilson – Cornell Lab of Ornithology

Scientific Advisors

Kevin McGowan – Cornell Lab of Ornithology
Kenneth Rosenberg – Cornell Lab of Ornithology
Christopher Wood – Cornell Lab of Ornithology

ISBN 978-0-9861782-0-7

9 780986 178207 >

CARTE DIEM PRESS
map the day

Dedication

This book is dedicated to the memory of Courtney R. Wilson, a talented and passionate young environmental scientist and educator whose star shone brightly and was diminished far too soon. With her love of learning and determination to make the world a better place, she inspired everyone who crossed her path.

Acknowledgments

Crossing Boundaries represents a collaborative effort among teachers, science educators, and scientists. We are grateful to the 60 teachers who participated in our National Science Foundation funded project and collaborated with us on all phases of curriculum design, field testing, and evaluation. This material was developed through the Crossing Boundaries Project (http://crossingboundaries.org) with support by the National Science Foundation under Grant No. 0833675. Any opinions, findings, and conclusions or recommendations expressed in this material are those of the authors and do not necessarily reflect the views of the National Science Foundation.

Birds Without Borders

Investigating Populations, Habitats, and Conservation of Birds in the U.S. and Abroad

Using citizen science and other data, students address questions such as why bird habitat requirements vary from species to species and why habitat protection for some must span continents. They consolidate their understanding by devising a plan for protection of a selected bird species or critical habitat in the U.S. or abroad.

Investigation 1. Discovering the Ecological Roles of Birds

Overview: Students identify and evaluate the arguments and claims in selected readings about ecosystem services, bird conservation and biodiversity.

Essential Questions

- What is meant by "ecosystem services"?

- What ecosystem services do birds provide?

- Why should we care about bird life, considering the ecosystem services that birds provide?

Technology: Videos, a word cloud generator, and an interactive online infographic

Investigation 2. Exploring Habitat Needs of Nesting Birds

Overview: Students use *NYS Breeding Bird Atlas* data and web resources to learn about the relationship between habitats and nesting bird distribution patterns.

Essential Questions

- How do environmental factors influence where birds live and breed?

- How are species adapted to specific environmental conditions?

- How do environmental factors and species-specific adaptations interact to determine where birds live and breed?

Technology: ArcGIS Online or Interactive PDF

Investigation 3. Determining Annual Life Cycles of Local Birds

Overview: Using eBird citizen science data and visualization tools, students discover what bird species live in their area year-round and which are seasonal residents, and they consider why some birds migrate while others do not.

Essential questions:

- Why do some birds migrate while others do not?

- How does this relate to each species' habitat requirements during nesting season and throughout the year?

- **Technology:** eBird web-based data visualization tools, *All About Birds* website

Investigation 4. Modeling Bird Population Trends

Overview: Using animated maps, students relate bird migration patterns to landscape features and habitat needs, and they consider the roles of citizen science and modeling in addressing these kinds of questions.

Essential question: What can modeling tell us about bird population dynamics?

Technology: eBird animated maps

Investigation 5. Tracking Birds with Citizen Science

Overview: Students use eBird citizen science data visualization tools to investigate the distribution of various bird species over time and space.

Essential question: How are citizen science data and modeling useful in tracking bird population dynamics?

Technology: eBird web-based data visualization tools, eBird website

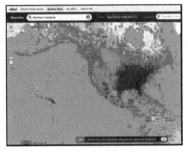

Investigation 6. Investigating Bird Biodiversity Across the Americas

Overview: Students create graphs and use maps, quantitative data, and web resources to investigate distribution of selected bird species in the U.S. and beyond.

Essential question: How does bird species richness compare between the U.S. and other countries in North and South America?

Technology: Excel (graphing)

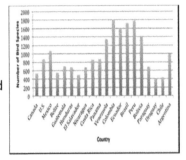

Investigation 7. Exploring Bird Conservation Needs at Home and Abroad

Overview: Using maps, quantitative data, and web resources, students investigate factors affecting the distribution of bird species across North and South America and consider the implications for conservation.

Essential questions:

- What are some reasons for the variation in distribution of bird species across North and South America?

- What countries or regions should be targeted for bird conservation?

- Why does conservation of some migratory species require international collaboration?

Technology: Excel (graphing), ArcGIS Online or Interactive PDF, State of the Birds website/report

Investigation 8. Creating a Conservation Plan

Overview: Students learn about needs and opportunities for bird conservation in the U.S. and neighboring countries. They propose an action plan to protect or restore a selected bird species or habitat, and they create a presentation to share with peers or the public.

Essential questions:

- What threats do birds face, in the U.S. and beyond our borders?

- What can be done to protect or restore one or more critical habitats or threatened species?

Technology: *State of the Birds* video, web-based presentation software such as Capzles

Contents

Investigation 6. Investigating Bird Biodiversity Across the Americas

Teacher

Investigation 7. Exploring Bird Conservation Needs at Home and Abroad

Teacher

Investigation 8. Creating a Conservation Plan

Teacher

Foreword

"It was true, she thought, that the big things awe us, but the little things touch us."

- Bess Streeter Aldrich, from *A White Bird Flying*

In many ways, the birding community represents the original citizen scientist community. Before the phrases *citizen science* or *volunteered geographic information* had entered the lexicon of the academy or the general public, before *Geographic Information Systems* came to be, those who observed birds kept meticulous records of the creatures they loved. Most of those records were tied to specific geographic locations. It is fitting, then, that this book focuses on citizen science and on the geographic aspects of birds.

It is also fitting that this book forms the first component of the series *Exploring Biodiversity Using Information Technologies,* for birds have long been an important part of the biodiversity of any ecoregion. Changes in the variety and populations of bird species have also served as key indicators of the impact of humans on biodiversity, whether that impact has been negative (as in the case of the passenger pigeon) or positive (such as with the passage of national wildlife refuge legislation).

For people through the centuries, observing or reading about birds has fostered environmental consciousness, the awareness not only of the interconnectedness of living things and their surroundings, but also of the beauty and fragility of life. Reading the book *The Last Great Auk* as an 11-year old provided my own watershed moment. Even though I knew that the book would end with the auk's extinction on Eldey Island in 1844, I said to myself "Can we prevent human-caused extinctions from happening in the future?" The manner in which the author (Eckert) made the last family of great auks so *personal,* so like *us,* only heightened my personal conviction to make my career in the environmental field. In part because of the direct threat to many species of wild birds in the late 19th Century, the environmental movement literally "took to flight," becoming a part of the international conversation.

Birds Without Borders isn't just a catchy title—themes of location, place, and space run throughout the book. Students using the book's techniques and methods learn concepts of geographic patterns, relationships, and trends. They understand how bird habitats are influenced by landforms, hydrology, climate, ecoregions, oceans, cities, and land use patterns, from local to global scales. They discover how conservation measures must also take into account these same factors.

This book and this series form a part of the quiet geotechnology revolution all around us. As nearly everything is becoming mapped and monitored, educators have an unprecedented opportunity to use geo-located data sets and tools in the field and online with their students. However, an equally important theme running through *Birds Without Borders* is that simply having access to technological tools does not guarantee adoption of those tools, nor does it guarantee transformation of educational practice. Indeed, being able to use the tools is not the end goal. Rather, it is the effective integration of these tools into active learning and inquiry that makes the investigations so powerful. The book also makes a strong case that making predictions, creating management plans, and articulating and communicating those plans cannot be *tacked onto* an investigation; they must be a *key part* of that investigation. Together, the inquiry process modeled in this book can enhance biology, technology, mathematics, environmental science, geography, and other disciplines. The book looks to an even greater goal—that students using these tools and methods transform society through the wise decisions that they make in their future careers.

One of my favorite things about the book and the Crossing Boundaries project is its focus on asking good questions. Thoughtful questions lead to rich investigations, additional questions, and critical thinking skills. The book, tied to the Next Generation Science Standards and Common Core standards, highlights writing, speaking, listening, and integrating knowledge and ideas. Non-fiction texts are included in authentic settings for students to discuss and use information. I appreciate the attention to international bird migration and conservation, supporting the "without borders" theme.

This book and series fit in well with a trend of books becoming *courseware.* Courseware includes discussion forums, implementation stories, data, web maps, tools, lessons, assessments, and more—in short, everything an educator or student needs. Thus, rather than a stand-alone resource that gets stale, the book forms a part of a living ecosystem that improves over time. In so doing, "book" is almost a misnomer for this resource. *Birds Without Borders* provides core content knowledge on concepts such as how ecosystems function and why birds matter. The book also provides practical lesson plans, hands-on activities, and resources, focusing unwaveringly on data. Through grappling with data stored in a variety of formats, students learn how to deal with using data and uncertainty in investigations. That practicing classroom educators were involved in creating and testing the activities is evident, because the activities are rich, engaging, and meaningful. They make use of web-based mapping tools and mobile technologies that have become incredibly powerful and yet easy to use, but low-tech options are always included.

Having served on the advisory board for the Crossing Boundaries project, having worked with Nancy Trautmann, Jim MaKinster, and having been fortunate to know Courtney Wilson while she was still with us, I know the deep impact that this project has had on educators. Through this book and others in the series, it gladdens me to know that this project will continue to inspire future educators and their students.

Why does all of this matter? Consider one more quote from Bess Streeter Aldrich from *A White Bird Flying*, "...For can you think how it would be, to never, never hear a meadow lark sing again...?"

- Joseph J. Kerski, Geographer, PhD., GISP
Esri Education Manager
University of Denver, Adjunct Instructor

Introduction

In *Birds Without Borders*, students analyze spatial and quantitative data, examine trends, make predictions, create management plans, and present and defend their results. Nonfiction texts and multimedia resources provide context and background for scientific exploration of real-world biodiversity data and issues in settings ranging from local to international.

This book is part of a series created through the Crossing Boundaries project. Originally funded by the National Science Foundation, Crossing Boundaries brings together education specialists, scientists, and teachers to create resources that engage students in investigations related to biodiversity. The project focuses on conservation science in the U.S. and abroad in order to increase students' awareness of ways in which they can apply scientific knowledge and skills to contribute toward positive solutions to environmental issues both at home and around the globe, http://www.crossingboundards.org.

Each Investigation poses one or more "essential questions" that trigger students' curiosity and highlight the relevance of key science concepts. With no easy answers, the questions are interdisciplinary in nature and can be addressed through multiple modes of inquiry.

Teachers were key partners in shaping, drafting, and piloting these Investigations. Drafts were reviewed and field-tested by middle and high school science teachers with students ranging from academically challenged to honors and advanced placement. Teacher reflections and classroom data documented positive outcomes for student learning, interest, and engagement at each of these levels and with all types of students.

Why Birds?

Wherever we live, birds share our environment. In habitats ranging from schoolyards to mountaintops, over 800 avian species live in the United States. Some remain in one place year-round while others migrate seasonally and have annual life cycles spanning countries or even continents. Birds provide compelling hooks to our local environment and links to foreign lands. They also provide essential ecosystem services, ranging from pollination and seed dispersal to garbage disposal and nutrient cycling.

Exploring the world of birds draws students into learning about habitats and adaptations. Comparing species, students see that some species are adapted to living under a broad range of environmental conditions while others have highly specialized habitat requirements. Studying migration, they learn about the complex life cycles of some species and the ways in which citizen science and modeling are creating exciting new ways to track bird population dynamics over space and time. Comparing statistics on bird species across countries, students discover the latitudinal gradient in species richness from the Equator toward the poles, and they consider the complexities inherent in categorizing biodiversity using either ecological or political boundaries. Finally, discovering the challenges faced by birds draws students into weighing options for conservation action and designing plans for preservation or restoration of selected species or habitats based on categories drawn up by the tri-national *Partners in Flight* organization. Using authentic scientific data and conservation strategy resources, students tackle crucial issues of real-world relevance.

Audience

Birds Without Borders can be used as a module in biology, environmental science, general science courses, AP biology, or any other applicable science elective. The student readings and activities have been successfully used in courses ranging from seventh grade through advanced placement. This broad range is possible with differing types of scaffolding and levels of sophistication expected in data analysis and application of the results.

Integrated science or environmental science commonly is taught as an introductory or basic level high school science course. *Birds Without Borders* works well in this setting because it uses thought-provoking hands-on activities and does not assume detailed prior knowledge. Our pilot testing has shown that students who are not accustomed to envisioning themselves as scientists gain motivation and self esteem when challenged to carry out authentic investigations of real-world issues.

For advanced classes, *Birds Without Borders* provides opportunities to expand students' understanding of complex concepts related to the interdependency between organisms and the environment and the complex interrelationships among environmental variables, habitats, and species adaptations.

Applying Technology

Crossing Boundaries strives to make science accessible through activities in which students use technological applications analogous to those used by professional scientists. Rather than learning about the technological tools *per se*, students learn to apply them in exploring and addressing current relevant environmental issues.

Students who conduct the activities in this book will use a variety of technological tools. Required applications and data resources are available here: http://crossingboundaries.org/bwb.php

In a world in which students are bombarded with online resources and communication opportunities, students using the Crossing Boundaries curriculum put such tools to academic use and see myriad ways in which to apply them toward scientific discovery and conservation. For example, students conducting *Birds Without Borders* activities use geographic information systems (GIS) to explore spatial data. In Investigation 2, they investigate the relationship between environmental features and the nesting distribution of individual bird species within a single state. In Investigation 5, they use spatial data on a much broader scale to explore relationships between bird species richness and environmental features along a north/south gradient across the continent. They graph and map bird species richness by country, and they interpret citizen science data and model outputs to track population distributions and migration patterns. Using online data and other web resources, students address questions such as why bird habitat requirements vary from species to species and why conservation efforts typically must span countries or even continents. In the final Investigation, they use web-based communication tools to create multimedia presentations to share with their peers.

Challenging Students and Meeting Standards

Developed and refined by a project team encompassing educators, scientists, and teachers, the *Birds Without Borders* Investigations are designed to address key aspects of the Next Generation Science Standards (http://www.nextgenscience.org/) and Common Core Standards (http://www.corestandards.org/).

The Next Generation Science Standards (NGSS) mandate that students operate at the intersection of science content and practice as they make connections across disciplinary lines. *Birds Without Borders* Investigations achieve this goal by engaging students in real-world issues that inherently require cross-disciplinary approaches to data analysis, interpretation, and synthesis while addressing core science concepts at middle and high school levels.

The Common Core State Standards mandate that students learn how to determine the central ideas or conclusions of a text and cite specific textual evidence to support analysis of science and technical texts. *Birds Without Borders* supports development of these skills through targeted activity sheets accompanying scientific readings. The Investigations also support development of mandated skills related to researching, building a case, and using technology to create a presentation.

Teacher Resources

We know that teachers need flexible resources; therefore we have created each investigation with some standard components to give educators all they need to successfully integrate the activities into their own curriculum.

Publicly available resources (websites, videos, etc.) for this unit are available at the Crossing Boundaries website. (http://crossingboundaries.org/bwb.php)

Go to the online resource portal at http://gisetc.com/bwb/ with password *ornithology* for all maps, student documents and teacher resources in digital format. You also can download all the related materials found at the public Crossing Boundaries website.

In each investigation, you will find a collection of teacher and student documents, ready to copy and use in any educational setting.

For the Teacher:

- Investigation Overview (an introduction of essential questions and general information related to the investigation)
- Before You Start (a list of time, preparation, technology and materials needed)
- Conducting the Investigation (your step by step guide to teaching the investigation)
- PowerPoint Slides (thumbnail versions of slides available for download)

For the Students:

- Readings
- Worksheets
- Answer Keys

Next Generation Science Standards

	Investigation 1	Investigation 2	Investigation 3	Investigation 4	Investigation 5	Investigation 6	Investigation 7	Investigation 8
Middle School								
Matter and Energy in Organisms and Ecosystems								
Analyze and interpret data to provide evidence for the effects of resource availability on organisms and populations of organisms in an ecosystem. (MS-LS2-1)		X		X		X	X	X
Ecosystems: Interactions, Energy, and Dynamics								
Construct an explanation that predicts patterns of interactions among organisms across multiple ecosystems. (MS-LS2-2)		X	X	X	X	X	X	X
Evaluate competing design solutions for maintaining biodiversity and ecosystem services. (MS-LS2-5)								X
Earth and Human Activity								
Construct an argument supported by evidence for how increases in human population and per-capita consumption of natural resources impact Earth's systems. (MS-ESS3-4)	X							X
Engineering Design								
Define the criteria and constraints of a design problem with sufficient precision to ensure a successful solution, taking into account relevant scientific principles and potential impacts on people and the natural environment that may limit possible solutions. (MS-ETS1-1)								X
High School								
Ecosystems: Interactions, Energy, and Dynamics								
Use mathematical representations to support and revise explanations based on evidence about factors affecting biodiversity and populations in ecosystems of different scales. (HS-LS2-2)			X		X	X	X	
Design, evaluate, and refine a solution for reducing the impacts of human activities on the environment and biodiversity. (HS-LS2-7)								X
Evaluate the evidence for the role of group behavior on individual and species' chances to survive and reproduce. (HS-LS2-8)			X				X	
Earth and Human Activity								
Evaluate or refine a technological solution that reduces impacts of human activities on natural systems. (HS-ESS3-4)								X
Engineering Design								
Design a solution to a complex real-world problem by breaking it down into smaller, more manageable problems that can be solved through engineering. (HS-ETS1-2)								X

Next Generation Science Standards: Scientific Practices

	Investigation 1	Investigation 2	Investigation 3	Investigation 4	Investigation 5	Investigation 6	Investigation 7	Investigation 8
Asking questions and defining problems	X	X	X	X	X	X	X	X
Developing and using models				X				
Planning and carrying out investigations		X	X	X	X	X	X	X
Analyzing and interpreting data		X	X	X	X	X	X	
Using mathematics and computational thinking					X	X	X	
Constructing explanations and designing solutions		X	X	X	X	X	X	X
Engaging in argument from evidence	X	X	X	X	X	X	X	X
Obtaining, evaluating, and communicating information	X	X	X	X	X	X	X	X

	Investigation 1	Investigation 2	Investigation 3	Investigation 4	Investigation 5	Investigation 6	Investigation 7	Investigation 8
Reading Science and Technical Subjects								
Key Ideas and Details								
Cite specific textual evidence to support analysis of science and technical texts, attending to the precise details of explanations or descriptions. (RST.9-10.1)	X	X	X	X		X		X
Determine the central ideas or conclusions of a text; trace the text's explanation or depiction of a complex process, phenomenon, or concept; provide an accurate summary of the text. (RST.9-10.2)	X		X	X		X		X
Follow precisely a complex multistep procedure when carrying out experiments, taking measurements, or performing technical tasks, attending to special cases or exceptions defined in the text. (RST.9-10.3)		X	X	X	X	X	X	
Craft and Structure								
Determine the meaning of symbols, key terms, and other domain-specific words and phrases as they are used in a specific scientific or technical context relevant to grades 9-10 texts and topics. (RST.9-10.4)		X	X	X	X	X	X	X
Integration of Knowledge and Ideas								
Translate quantitative or technical information expressed in words in a text into visual form (e.g., a table or chart) and translate information expressed visually or mathematically (e.g., in an equation) into words. (RST.9-10.7)	X		X	X	X	X	X	
Compare and contrast findings presented in a text to those from other sources (including their own experiments), noting when the findings support or contradict previous explanations or accounts. (RST.9-10.9)		X	X	X	X	X		
Range of Reading and Level of Text Complexity								
By the end of grade 10, read and comprehend science/technical texts in the grades 9-10 text complexity band independently and proficiently. (RST.9-10.10)	X	X	X	X		X		X

	Investigation 1	Investigation 2	Investigation 3	Investigation 4	Investigation 5	Investigation 6	Investigation 7	Investigation 8
Writing								
Production and Distribution of Writing								
Use technology, including the Internet, to produce, publish, and update individual or shared writing products, taking advantage of technology's capacity to link to other information and to display information flexibly and dynamically. (W.9-10.6)								X
Research to Build and Present Knowledge								
Conduct short as well as more sustained research projects to answer a question (including a self-generated question) or solve a problem; narrow or broaden the inquiry when appropriate; synthesize multiple sources on the subject, demonstrating understanding of the subject under investigation. (W.9-10.7)		X	X	X	X	X	X	X
Gather relevant information from multiple authoritative print and digital sources, using advanced searches effectively; assess the usefulness of each source in answering the research question; integrate information into the text selectively to maintain the flow of ideas, avoiding plagiarism and following a standard format for citation. (W.9-10.8)		X	X		X		X	X
Draw evidence from literary or informational texts to support analysis, reflection, and research. (W.9-10.9)	X	X	X	X	X			X
Speaking and Listening								
Comprehension and Collaboration								
Initiate and participate effectively in a range of collaborative discussions (one-on-one, in groups, and teacher-led) with diverse partners on grades 9-10 topics, texts, and issues, building on others' ideas and expressing their own clearly and persuasively. (SL.9-10.1)	X	X	X	X	X		X	X
Integrate multiple sources of information presented in diverse media or formats (e.g., visually, quantitatively, orally) evaluating the credibility and accuracy of each source. (SL.9-10.2)		X	X	X	X		X	X
Presentation of Knowledge and Ideas								
Present information, findings, and supporting evidence clearly, concisely, and logically such that listeners can follow the line of reasoning and the organization, development, substance, and style are appropriate to purpose, audience, and task. (SL.9-10.4)								X
Make strategic use of digital media (e.g., textual, graphical, audio, visual, and interactive elements) in presentations to enhance understanding of findings, reasoning, and evidence and to add interest. (SL.9-10.5)								X

1 investigation

Discovering the Ecological Roles of Birds

Investigation 1 Photo Credits

Kelly Colgan Azar. Magnolia Warbler. (https://flic.kr/p/aimqEE)

Kelly Colgan Azar. Downy Woodpecker. (https://flic.kr/p/atLcYx)

Taza Schaming. Clark's Nutcracker. Cornell Lab of Ornithology

Investigation 1
Discovering the Ecological Roles of Birds

This Investigation builds understanding of the ecological roles played by birds, addressing the overarching question, "Why care about birds?" within the context of essential ecosystem services that they provide. Students identify and evaluate arguments and claims in selected videos and readings about ecosystem services, biodiversity, and birds to learn about crucial ecological interactions within any ecosystem and the environmental consequences that can occur when bird populations are disrupted.

Essential Questions

- What is meant by "ecosystem services"?

- What ecosystem services do birds provide?

- Why should we care about bird life, considering the ecosystem services that birds provide?

Science Concepts and Topics

This Investigation introduces the concept of **ecosystem services**, the natural ecological functions and processes that sustain human life and benefit society. Ecosystem services provided by birds include control of insects and rodents, dispersal of seeds, pollination of flowers, disposal of carrion, and cycling of nutrients. The importance of such services is captured in this quote:

> "Birds keep our ecosystems healthy, controlling pests and disease vectors by consuming immense quantities of insects and rodents, facilitating decomposition and nutrient cycling through the consumption of carrion, pollinating flowers, and dispersing seeds. They also excavate cavities and burrows essential for other wildlife. As birds migrate across the continent, they carry these services with them. The enormous number of shared landbirds can consume at least 100,000 metric tons of invertebrates daily (equivalent in weight to more than 20,000 African elephants!). Birds in Canada's boreal forest alone are estimated to provide $5.4 billion in pest control services each year." (*Saving Our Shared Birds*, p. 6)

Maintaining essential ecosystem functions such as these depends on protection of **biodiversity**, which is defined in terms of three components: the number of species living in a specified area, the genetic diversity within each species, and the variety of habitats represented. Ecosystem stability and function are determined by these three components of biodiversity. High levels of biodiversity tend to make an ecosystem relatively stable and resilient to environmental change.

Due to growing human populations, increasing resources consumption, and changing climate, **our birds are at risk**. Across the Western Hemisphere, habitat destruction and other challenges are reducing the numbers of individuals within species and placing ever-growing numbers of species at risk of extinction.

Maintaining both the diversity and abundance of birds is critical to their ongoing provision of essential ecosystem services. For migratory species, efforts must extend beyond U.S. borders to protect habitats and lives in our neighboring countries and across the continent.

Standards Addressed in This Investigation

Next Generation Science Standards – Middle School

- Construct an argument supported by evidence for how increases in human population and per-capita consumption of natural resources impact Earth's systems. (MS-ESS3-4)

Next Generation Science Standards – Scientific Practices

- Asking questions and defining problems
- Engaging in argument from evidence
- Obtaining, evaluating, and communicating information

Common Core (9-10th grade)

Reading Science & Technical Subjects

Key Ideas and Details

- Cite specific textual evidence to support analysis of science and technical texts, attending to the precise details of explanations or descriptions. (RST.9-10.1)
- Determine the central ideas or conclusions of a text; trace the text's explanation or depiction of a complex process, phenomenon, or concept; provide an accurate summary of the text. (RST.9-10.2)

Integration of Knowledge and Ideas

- Translate quantitative or technical information expressed in words in a text into visual form (e.g., a table or chart) and translate information expressed visually or mathematically (e.g., in an equation) into words. (RST.9-10.7)

Range of Reading and Level of Text Complexity

- By the end of grade 10, read and comprehend science/technical texts in the grades 9-10 text complexity band independently and proficiently. (RST.9-10.10)

Writing

Research to Build and Present Knowledge

- Draw evidence from literary or informational texts to support analysis, reflection, and research. (W.9-10.9)

Speaking and Listening

Comprehension and Collaboration

- Initiate and participate effectively in a range of collaborative discussions (one-on-one, in groups, and teacher-led) with diverse partners on grades 9-10 topics, texts, and issues, building on others' ideas and expressing their own clearly and persuasively. (SL.9-10.1)

Discovering the Ecological Roles of Birds
Before You Start...

CROSSING boundaries

Time

75-90 minutes, depending on what you ask of students in terms of the readings this Investigation.

Preparation

Download the videos or check to make sure you will be able to stream them from the web in your classroom.

- Make copies of the Student Readings and Worksheets.
- Select a website for creating a word cloud (see Step 3b).

Technology

A-V Equipment

- LCD Projector
- Screen or Whiteboard

Computers

- Acrobat Reader
- Internet connection

Information and Communication Technology

- Videos, a web-based word cloud generator, and an interactive online infographic

Low-Tech Options

- Use the "Why Birds Matter" poster in place of the interactive online version of this poster.

Materials

Videos

- Marita Davison – In the Field
- Taza Schaming – In the Field

(Stream or download videos— http://crossingboundaries.org/bwb.php)

Student Readings

- Ecosystem Services
- Guam's Birds Gone: Can Forests Survive?
- How Birds Keep our World Safe from Plagues of Insects
- Why Birds Matter – use the provided poster if the interactive infographic here will not work (Note: the interactive uses Flash so it will not work on iPads) http://crossingboundaries.org/bwb.php

Worksheets

- Ecosystem Services
- Guam's Birds Gone
- How Birds Keep Our World Safe from Plagues of Insects

Discovering the Ecological Roles of Birds
The Investigation

Investigation Overview

1. Learn about bird research through the eyes of two Crossing Boundaries Conservation Scientists.

2. Read and reflect about ecosystem services.

3. Create a word cloud illustrating ecosystem services provided by birds.

4. Reflect on why birds matter.

Learning Objectives

Students will be able to:

- Define and understand the meaning of ecosystem services and related terms.

- Explain the ecological roles played by birds.

- Describe the importance of bird conservation in terms of ecosystem services provided by birds.

Key Concepts

- Biodiversity
- Species richness
- Conservation
- Ecosystem services
- Threats to biodiversity

Technology Overview

- This Investigation uses videos, a web-based word cloud generator, and an interactive web-based infographic.

Conducting the Investigation

1. Learn about bird research through the eyes of two Crossing Boundaries Conservation Scientists.

   a. Explain that you will be watching a brief video about a scientist named Marita, and prompt students to think while they are watching about why she is interested in excluding flamingos from parts of the lakes in her study.

   b. As a class, view Marita Davison's *In the Field* video (~7 minutes).

   c. Discuss the purpose and methods of Marita's research.

   Marita's research explores how human activities affect individual species and the ecosystems in which they live, focusing on flamingos in lakes in the Andes mountains of Bolivia. By fencing flamingos out of some sections of these lakes, she can measure changes in primary productivity and other ecosystem processes. Her goal is to provide a better understanding of what would happen in these lakes if flamingos were to become extinct.

   d. As a class, watch Taza Schaming's *In the Field* video (~12 minutes). Before starting the video, prompt students to consider while watching why Taza is interested in tracking Clark's Nutcrackers in the Wyoming wilderness.

Marita Davison: *In the Field*

Taza Schaming: *In the Field*

e. Discuss the purpose of Taza's research.

Taza is tracking these birds to learn what they eat, how much they move around to collect food, and how they relate to each other socially. She wants to understand the interdependencies between Nutcrackers and Whitebark Pine trees, which are dying because of a beetle epidemic and infections from blister rust, an invasive pathogen. The nutcrackers eat seeds from Whitebark Pines, and the trees rely on the birds to plant their seeds. Taza aims to identify conservation management actions that could help to maintain a healthy ecosystem and make it possible for Clark's Nutcrackers to survive in this region in the midst of environmental change.

2. Read and reflect about ecosystem services.

a. Have students read **Ecosystem Services** and respond to the questions in the **Ecosystem Services Reading Questions Worksheet**.

b. Discuss the worksheet.

1) How did you define ecosystem services? *Definition: The processes and resources supplied by ecosystems that are of benefit to humans.*

2) What is a service from the ecosystem in which you live that you receive on a day-to-day basis yet fail to pay for? *Examples cited in the article include flood protection, natural medicinal products, pollination, and water purification, but other ecosystem services also are acceptable responses here.*

3) How much do you think the ecosystem service you identified is worth in terms of a fee people might have to pay? *Depends on response to #2 and should include some sort of calculation of cost for a product or service.*

4) What is one example you identified where a human activity disrupts ecosystem services in your own community? *Could include pollution, introduction of non-native species, overharvesting of fisheries, destruction of wetlands, soil erosion, deforestation, urban sprawl. Student responses should include connections to local issues.*

5) What ecosystem services provided by birds are addressed in Marita research? *Marita's research addresses the influence of flamingos on primary productivity and food webs in lakes in the Bolivian Andes. Relevant ecosystem services include nutrient cycling and maintaining biodiversity. By excluding flamingos from some areas, she can measure their influence on the diversity and amount of algae that form the basis for the lakes' food webs.*

6) What ecosystem services provided by birds are addressed Taza's research? *Taza is studying the relationship between Clark's Nutcrackers and Whitebark Pine trees. Seed dispersal is one ecosystem service that she discusses. Another is nutrient cycling. Each nutcracker collects and buries up to 35,000 pine seeds per year, and those not eaten can germinate and grow into new trees.*

c. Break students into groups to read either *Guam's Birds Gone: Can the Forest Survive?* or *How Birds Keep Our World Safe from Plagues of Insects* and fill in the accompanying worksheets.

d. As a class, integrate concepts from the readings by discussing the ecological roles played by birds.

   1) What ecosystem service did birds provide on Guam? *Seed dispersal – when birds eat fruit, they move seeds and affect the distribution of the fruiting trees.*

   2) Why is seed dispersal so important in the survival of a species, especially for the fruit trees on Guam? *Seeds that sprout directly below the parent tree appear to be more susceptible to fungal infections and more likely to be killed by predators than those that have been dispersed more broadly.*

   3) What caused the disappearance of Guam's birds? *Introduction of a non-native species, the brown tree snake, which eats birds. Previously no snakes existed on Guam, and the birds were not adapted to have defenses against this predator.*

   4) What ecosystem service is discussed in the insect article? *Pest control. Birds eat insects and reduce outbreaks that could otherwise destroy forests or crops.*

   5) What are some ways in which we can help birds? *Possibilities from the article include:*

      - *Providing next boxes for cavity-nesting birds*

      - *Refraining from chopping down dead trees, which provide needed habitat for woodpeckers and other cavity-nesting birds*

      - *Integrated Pest Management (IPM), including providing habitat that supports birds and helps them to control insect pests*

3. Create a word cloud illustrating ecosystem services provided by birds.

   a. Have each student brainstorm a list of ecosystem services provided by birds.

   b. Create a class word cloud portraying these lists by having students take turns entering their items into the text box and then generate your word cloud using a site such as http://www.wordle.net/create or http://worditout.com/word-cloud/make-a-new-one. *The greater the size of each word, the more times it has been listed by students. Possibilities include pollinating flowers, dispersing seeds, controlling pests such as insects and rodents, recycling wastes such as carrion, dispersing nutrients in the form of guano, eating and being eaten as part of the web of life. Cultural services also are commonly included. These are the benefits humans derive from natural ecosystems through recreation, aesthetic and spiritual appreciation, education, etc.*

Sample Word Cloud

4. Reflect on why birds matter.

   a. Using the **Why Birds Matter** online infographic (http://crossingboundaries.org/bwb.php) or poster, have students explore the highlighted bird species and determine whether they can add any additional services to the collection they made in Step 3.

   b. To conclude, ask students to write a quote to inspire and inform people about why birds matter. For reference, they could use ideas and facts from the article "Why Do Birds Matter?" at http://crossingboundaries.org/bwb.php.

   c. Assemble the students' quotes on a poster in your classroom or a webpage to display on your school's site.

   d. Explain that in following classes you will be exploring various aspects of bird life in the U.S. and across our continent and considering implications for conservation of threatened species and habitats.

   e. Possible extension: Consider ways to assess the economic value of ecosystem services and think of examples of how such processes could be applied to addressing the question of why it is important to maintain vibrant and diverse natural bird populations (see **What Are Species Worth? Putting a Price on Biodiversity**, http://e360.yale.edu/feature/what_are_species_worth_putting_a_price_on_biodiversity/2322).

Why Birds Matter infographic

Ecosystem Services
Student Reading

Have you ever considered that the cereal you eat is brought to you each morning by the wind, or that the glass of clear, cold, clean water drawn from your faucet may have been purified for you by a wetland or perhaps the root system of an entire forest? Trees in your front yard work to trap dust, dirt, and harmful gases from the air you breathe. The bright fire of oak logs you light to keep warm on cold nights and the medicine you take to ease the pain of an ailment come to you from Nature's warehouse of services. Natural ecosystems perform fundamental life-support services upon which human civilization depends. Unless human activities are carefully planned and managed, valuable ecosystems will continue to be impaired or destroyed.

What are ecosystem services?

Ecosystem services are the processes by which the environment produces resources that we often take for granted such as clean water, timber, and habitat for fisheries, and pollination of native and agricultural plants. Whether we find ourselves in the city or a rural area, the ecosystems in which humans live provide goods and services that are very familiar to us.

Ecosystems provide "services" that:

- moderate weather extremes and their impacts
- disperse seeds
- mitigate drought and floods
- protect people from the sun's harmful ultraviolet rays
- cycle and move nutrients
- protect stream and river channels and coastal shores from erosion
- detoxify and decompose wastes
- control agricultural pests
- maintain biodiversity
- generate and preserve soils and renew their fertility
- contribute to climate stability
- purify the air and water
- regulate disease carrying organisms
- pollinate crops and natural vegetation

What is an ecosystem?

An ecosystem is a community of animals and plants interacting with one another and with their physical environment. Ecosystems include physical and chemical components, such as soils, water, and nutrients that support the organisms living within them. These organisms may range from large animals and plants to microscopic bacteria. Ecosystems include the interactions among all organisms in a given habitat. People are part of ecosystems. The health and well-being of human populations depends upon the services provided by the ecosystems and their components - organisms, soil, water, and nutrients.

What are ecosystem services worth?

Natural ecosystems and the plants and animals within them provide humans with services that would be very difficult to duplicate. While it is often impossible to place an accurate monetary amount on ecosystem services, we can calculate some of the financial values. Many of these services are performed seemingly for "free", yet are worth many trillions of dollars, for example:

- Much of the Mississippi River Valley's natural flood protection services were destroyed when adjacent wetlands were drained and channels altered. As a result, the 1993 floods resulted in property damages estimated at twelve billion dollars partially from the inability of the Valley to lessen the impacts of the high volumes of water.

- Eighty percent of the world's population relies upon natural medicinal products. Of the top 150 prescription drugs used in the U.S., 118 originate from natural sources: 74 percent from plants, 18 percent from fungi, 5 percent from bacteria, and 3 percent from one vertebrate (snake species). Nine of the top 10 drugs originate from natural plant products.

- Over 100,000 different animal species - including bats, bees, flies, moths, beetles, birds, and butterflies - provide free pollination services. One third of human food comes from plants pollinated by wild pollinators. The value of pollination services from wild pollinators in the U.S. alone is estimated at four to six billion dollars per year.

Before it became overwhelmed by agricultural and sewage runoff, the watershed of the Catskill Mountains provided New York City with water ranked among the best in the nation by Consumer Reports. When the water fell below quality standards, the city investigated what it would cost to install an artificial filtration plant. The estimated price tag for this new facility was six to eight billion dollars, plus annual operating costs of 300 million dollars - a high price to play for what once was free. New York City decided instead to invest a fraction of that cost ($660M) in restoring the natural capital it had in the Catskill's watershed. In 1997, the city raised an Environmental Bond Issue and is currently using the funds to purchase land and halt development in the watershed, to compensate property owners for development restrictions on the land, and to subsidize the improvement of septic systems.

> The choices we make today in how we use land and water resources will have enormous consequences on the future sustainability of earth's ecosystems and the services they provide.

How are ecosystem services "cut off"?

Ecosystem services are so fundamental to life that they are easy to take for granted and so large in scale that it is hard to imagine that human activities could destroy them. Nevertheless, ecosystem services are severely threatened through (1) growth in the scale of human enterprise (population size, per-capita consumption, and effects of technologies to produce goods for consumption) and (2) a mismatch between short-term needs and long-term societal well-being.

Many human activities disrupt, impair, or reengineer ecosystems every day including:

- runoff of pesticides, fertilizers, and animal wastes
- pollution of land, water, and air resources
- introduction of non-native species
- overharvesting fisheries
- destruction of wetlands
- erosion of soils
- deforestation
- urban sprawl

Ecology and ecosystem services

Ecologists work to help us understand the interconnection and interdependence of the many plant and animal communities within ecosystems. Although substantial understanding of many ecosystem services and the scientific principles underlying them already exists, there is still much to learn. The tradeoffs among different services within an ecosystem, the role of biodiversity in maintaining services, and the effects of long and short-term perturbations are just some of the questions that need to be further explored. The answers to such questions will provide information critical to the development of management strategies that will protect ecosystems and help maintain the provisions of the services upon which we depend.

For More Information

Issues in Ecology, "Ecosystem Services: Benefits Supplied to Human Societies by Natural Ecosystems, No. 2, Spring, 1997, Ecological Society of America. Available on ESA's web site at http://www.esa.org/esa/wp-content/uploads/2013/03/issue2.pdf.

Nature's Services, Societal Dependence on Natural Ecosystems, Gretchen C. Daily, Editor, Island Press, 1997.

Teaming with Life: Investing in Science to Understand and Use America's Living Capital, President's Committee of Advisors on Science and Technology, March 5, 1998. Available through the Internet at http://www.whitehouse.gov/WH/EOP/OSTP/html/OSTP_Home.html.

esa

Prepared by the

Ecological Society of America
1900 M Street, NW, Suite 700,
Washington, DC 20036

Name: _____ Period: _____ Date: _____

Ecosystem Services
Worksheet

1. In your own words, define "ecosystem services" and give an example.

    Definition:

    Example:

2. In the reading, monetary values are given to several ecosystem services. What is a service that you receive on a day-to-day basis yet fail to pay for?

3. If you could assign a fee, what do you feel would be a realistic amount to pay? (Justify your response by calculating the cost of something that you or your family pays for.)

4. Select a human activity that disrupts ecosystem services and describe how it applies in your community.

5. Choose either Marita's or Taza's research and identify which ecosystem services she addresses in her video.

Name: _____ Period: _____ Date: _____

Ecosystem Services
Worksheet Answer Key

1. In your own words, define "ecosystem services" and give an example.

   Definition:

   The processes and resources supplied by ecosystems that are of benefit to humans.

   Example:

   Many possibilities. Could include purifying air and water, pollination, decomposition of wastes, nutrient cycling, providing timber or other resources, moderating weather extremes.

2. In the reading, monetary values are given to several ecosystem services. What is a service that you receive on a day-to-day basis yet fail to pay for?

   Examples cited in the article include flood protection, natural medicinal products, pollination, and water purification, but any ecosystem services are acceptable responses here.

3. If you could assign a fee, what do you feel would be a realistic amount to pay? (Justify your response by calculating the cost of something that you or your family pays for.)

   Depends on response to #2, should include some sort of calculation of cost for a product or service.

4. Select a human activity that disrupts ecosystem services and describe how it applies in your community.

   Could include pollution, introduction of non-native species, overharvesting of fisheries, destruction of wetlands, soil erosion, deforestation, urban sprawl. Student response should include connection to a local issue.

5. Choose either Marita's or Taza's research and identify which ecosystem services she addresses in her video.

   Marita's research addresses the influence of flamingos on primary productivity and food webs in lakes in the Bolivian Andes. Relevant ecosystem services include nutrient cycling and maintaining biodiversity. By excluding flamingos from some areas, she can measure their influence on the diversity and amount of algae that form the basis for the lakes' food webs.

   Taza is studying the relationship between Clark's Nutcrackers and Whitebark Pine trees. Seed dispersal is one ecosystem service that she discusses. Another is nutrient cycling. Each nutcracker collects and buries up to 35,000 pine seeds per year, and those not eaten can germinate and grow into new trees.

Guam's Birds Gone:
Can Forest Survive?

By Haldre Rogers, 23 January 2009

Student Reading

Can forests that have lost all of their birds still function normally? This is an important question for the now bird-less forests on the island of Guam, an island in the western Pacific.

How did Guam lose its birds? In the mid-1940s, the brown tree snake was accidentally introduced to what was then snake-free Guam. This snake became Guam's new top predator and ate its way through a buffet of the island's bird community. As a result, 10 of the island's 12 forest bird species are now extinct on Guam and the two surviving forest bird species remain only in tiny, localized populations where snakes are controlled. Guam's now silent forests currently hold about 13,000 snakes per square mile.

I started to think about the potential ecological impacts of bird loss in 2002, when two years out of college, I was hired by the U.S. Geological Survey to develop a "Rapid Response Team" that would identify and eradicate new populations of brown tree snakes on U.S.-associated Pacific Islands. Although I had heard the snake story in my college conservation biology course, I did not know where Guam was when I applied for the job. Yet, three weeks later, I was on a plane headed there.

Bird loss and seed movement

As I worked on Guam during the next three years, I often wondered why no one was studying how Guam's bird losses impacted the forests' remaining organisms. So in 2005, I began a Ph.D. program in biology at the University of Washington to investigate how bird loss changes the movement of seeds around Guam's forests.

This spring, I, along with my co-advisers Joshua Tewksbury and Janneke Hille Ris Lambers, our collaborator at the University of Guam Ross Miller, and my field assistant Theresa Feeley-Summer, began to examine whether the loss of birds had caused changes in how the seeds they typically eat are distributed.

The study is funded by the Budweiser Conservation Scholarship and the National Science Foundation (NSF) Integrated Graduate Education, Research, and Training (IGERT) and Graduate Research Fellowships. In this study, we set seed traps at various distances from fruiting False Elder (Premna obtusifolia) trees in the forests of Guam and Saipan, a nearby island with birds, and then counted the number of seeds that fell into each trap. This shows us how far the seeds of fruiting trees are traveling in Guam's bird-less forests as compared to Saipan's forests with birds.

Screen-door netting and a mile of PVC

The first step in our research was to design traps to catch falling seeds using locally available materials. This task required many trips to the new Home Depot on Guam; we purchased the store's entire supply of screen-door netting, flexible PVC piping and PVC connectors. Believe me, I got some strange looks when I asked the Home Depot sales person for 2,000 feet of screen door netting and a mile of PVC.

Although our study is still ongoing, we have already produced some important results: we found that all of the seeds from the fruiting trees on Guam remained near their parent trees and maintained intact seed coats. By contrast, many more of the seeds from the fruiting trees on Saipan were found without seed coats away from their parent tree.

The differences between the distributions of the seeds on Guam and Saipan can be attributed to the differences in their bird populations: In Saipan's forests, birds stop at fruiting trees, eat fruit, swallow the seeds and then fly to the next tree, where they defecate, effectively moving seeds away from where they are produced. We believe the handling of seeds by birds removes the seed coat and promotes the germination of seeds. In the bird-less forests of Guam, however, fruits ripen, fall off of the tree and settle at tree bases without being eaten or moved by birds.

Unfortunately, our results do not bode well for the

future of Guam's fruit-producing trees. Research from around the world has shown that seeds from fruits falling under parent trees (like fruits in Guam) tend to experience higher mortality from predators and fungal infections than seeds that are moved away from their parent trees. In addition, for many species, seeds that are not handled by birds are less likely to germinate than seeds that are handled by birds.

More on seed dispersal

Where will our research go from here? I hope that Home Depot has restocked its screen-door netting and PVC piping, because we plan to construct about 1,000 more seed traps. We will use them to study seed dispersal for 14 more species of trees. This will give us a community perspective on seed dispersal patterns.

We will also investigate how the lack of bird handling and seed dispersal by birds impacts the germination and growth rate of seeds. In addition, we will evaluate the impacts of bird loss on local people by interviewing people who extract forest products for traditional uses. Several tree species used for medicinal purposes or as carving wood have seeds dispersed by birds, and thus may be experiencing population declines noticed by local people.

Although the introduction of a non-native snake caused Guam's bird loss, other factors are causing bird losses in forests around the world. The ecological impacts of all of these declines — no matter what the cause — are likely to be similar. Therefore, the complete loss of Guam's birds provides an extreme example that can inform us about the ecosystem impacts of bird losses around the world. The results of our research may be used by conservationists to develop and apply timely management approaches that will minimize the ecological impacts of bird loss.

A video of the researcher setting seed traps is posted https://www.youtube.com/watch?v=f6zKCngC2ko.

This article is made available by the National Science Foundation:

http://www.nsf.gov/discoveries/disc_summ.jsp?cntn_id=114051&org=DEB

Name: _____ Period: _____ Date: _____

Guam's Birds Gone
Worksheet

1. What was the primary cause for the loss of bird populations on Guam?

2. What "ecosystem service" is affected by the loss of birds on Guam?

3. What is helping Guam's two remaining species of birds to survive?

4. Why is seed dispersal so important in the survival of a species, especially for the fruit trees mentioned in the reading?

5. How are these scientists studying the effects of birds on seed dispersal on Guam, now that most of Guam's birds are gone?

6. Hypothesize what other factors are causing loss of bird diversity throughout the world.

CROSSING
boundaries

Name: _____ Period: _____ Date: _____

Guam's Birds Gone
Worksheet Answer Key

1. What was the primary cause for the loss of bird populations on Guam?

   Introduction of a non-native species, the brown tree snake, which eats birds. Previously no snakes existed on Guam, and the birds were not adapted to have defenses against this predator.

2. What "ecosystem service" is affected by the loss of birds on Guam?

   Seed dispersal – when birds eat fruit, they move seeds and affect the distribution of the fruiting trees.

3. What is helping Guam's two remaining species of birds to survive?

   They are surviving only in areas where people are keeping the snake population under control.

4. Why is seed dispersal so important in the survival of a species, especially for the fruit trees mentioned in the reading?

   Seeds that sprout directly below the parent tree appear to be more susceptible to fungal infections and more likely to be killed by predators than those that have been dispersed more broadly.

5. How are these scientists studying the effects of birds on seed dispersal on Guam, now that most of Guam's birds are gone?

   They are comparing seed dispersal on Guam (without birds) to dispersal of the same species of tree seeds on a similar nearby island where birds are present. On both islands, they set seed traps at various distances from the fruiting trees and then count the number of seeds that fall into each trap.

6. Hypothesize what other factors are causing loss of bird diversity throughout the world.

   This information is not included in the article. Possibilities could include loss of habitat, competition with invasive species, over-hunting, and loss of food due to climate change or other environmental factors.

How Birds Keep our World Safe from the Plagues of Insects

Fact Sheet by the Smithsonian Migratory Bird Center

Student Reading

The Life of an Insect

Several species of insects including the Western Spruce Budworm, Gypsy Moth, Western Pine Beetle, and the Eastern Spruce Budworm experience population cycles in which populations remain low for several years and are followed by outbreaks (population explosions). During non-outbreak years, these insects are usually confined to small areas where trees are subject to adverse conditions, such as drought, and are too weak to defend against the insects.

Population outbreaks of some insect species can have a devastating effect on the forest because the insects severely defoliate the trees or attack the bark. Vast areas of forest have been killed during outbreaks in the past.

The basic life cycle of outbreak insects is a rapid growth of the larvae (caterpillars) during a short period, usually in June to mid-July, then a pupae stage (cocoons) in which the larvae change into adults (moths, butterflies, beetles), and finally the adult stage in which breeding and egg-laying take place. In some species the pupae stage will last through the winter, in others, the adults emerge in the same summer.

Dodging Death

Insects are subject to a myriad of threats including adverse weather, disease, parasites, habitat destruction, insecticides, and predation from spiders, ants, beetles, mammals, reptiles, amphibians, and birds. In the face of these threats, insects have evolved with complex methods for survival.

Predator-avoidance strategies are as varied as the insects themselves. Some create poisonous chemicals in their bodies, while others may have spines. Caterpillars and pupae often match their surroundings' color patterns, and some even mimic the shapes of leaves or twigs. Other species hide in dead, curled leaves, on the undersides of green leaves, in crevices in bark, under leaf litter on the ground, or in flowers. Some have even evolved feeding patterns to avoid predators, such as feeding at night, foraging in hidden spots, or by living and feeding under the bark. Others snip off the leaves that they fed on during the day in an attempt to trick birds that search for them on partially eaten leaves.

Birds Kill Bugs Dead

For all of the tactics insects have developed to avoid predation, they still face many species of birds that are highly adapted, consummate insect-eaters:

- Outbreak insects are often infected with parasites. Many birds can identify the infected insects, and often choose to eat those that are not parasitized. By preying only on healthy individuals, birds greatly add to the effect of parasites in reducing insect populations.

- Birds can spread viral infections among the insect pests. By eating beetles and their viruses and by defecating these viruses along tree trunks, birds inadvertently spread it to bark beetles in the same tree and throughout the forest.

- The breeding season for birds occurs when the insect populations are their highest. During insect outbreaks, some birds will increase the number of offspring that they raise to take advantage of the abundant food supply.

- Birds are highly mobile and many species of birds will take advantage of a local insect outbreak by moving into the infected area. Some of these invasions can increase the normal numbers of birds in an area by 80 times.

- Birds like to feed large, juicy insects to their young. Relatively few insects survive past the egg and small, young larval stages. By feeding on large, late stages of caterpillars, and on pupae and adults, birds become a key force in depleting insect populations.

- Birds can alter their diets to feed almost exclusively on an insect pest during an outbreak, if it becomes profitable for them to do so. They can develop a search image for this new prey and can learn how to hunt for it more efficiently. Factors that help determine which insects birds select as prey are; insect density, body size and nutritional content, ease of capture, palatability (presence of chemical defenses or parasites), and density of potential competitors (other birds, mammals, ants, spiders, and predacious insects).

- Along with developing a search image, birds can change their foraging locations and foraging behavior in response to an insect outbreak. When a vast quantity of insects is found in the canopy of trees, many ground or shrub-dwelling birds may ascend into the canopy to feed. Similarly, during a hatch of flying insects, birds that generally feed by plucking caterpillars off leaves may instead fly after the insects and capture them in mid-air.

- Some foraging strategies of birds can alter an insect species' preferred habitat to such an extent that it kills many of those insects. For instance, by flaking bark off tree trunks, woodpeckers will expose bark beetles to temperature extremes, loss of moisture, parasites, and predators, all of which result in increased deaths.

- Birds can affect the evolution of insects by increasing the cost of avoidance strategies to insects. Many of the adaptations can decrease the insect's efficiency in feeding and/or ability to lay the greatest, potential number of eggs.

Battling the Bugs

Bird predation may play a critical role in reducing and/or maintaining low populations of insect prey during non-outbreak years and in significantly increasing the time between outbreaks. Studies have shown that birds can eat up to 98% of budworms and as much as 40% of non-outbreak species in eastern forests and can alter the population cycles and lower the population peaks when an outbreak does occur.

Increased numbers of birds in patches of forest with high insect pest density during a non-outbreak year may result in the elimination of those insects, and can alter the location and spread of a subsequent outbreak.

Orchards near woodlots tend to have a higher number of birds which result in a higher predation rate of agricultural insect-pests. In some orchards, birds eat up to 98% of the over-wintering Codling Moths, and can successfully control the pest population.

Helping Birds Help Us

There is much that we can do to promote the effectiveness of birds as predators of harmful insect, thereby helping ourselves financially and environmentally. For example, we can encourage birds to take up residence in an area. Purple Martins have long been known as one of the most affective mosquito repellents. Protecting an existing colony, or helping the establishment of one is an important management tool.

In Europe, there have been numerous, successful, programs to provide nest boxes for cavity-nesting birds such as the Pied Flycatcher. These birds can substantially reduce the insect pest population without the economic, health and environmental costs of pesticides.

Managing for snags in a forest or woodlot can greatly increase the number of woodpeckers and other cavity-nesting birds. These species are highly efficient predators of insects, and can have a marked effect on insect populations.

One of the most promising forms of insect control is Integrated Pest Management (IPM), in which birds play a key role. The success and economic feasibility of these programs may depend on the number and diversity of birds in an area. Providing hedgerows, woodlots, streamside habitat, and shade trees in an agricultural landscape can provide cover and nesting areas for birds. Insect outbreaks can annually destroy hundreds of millions of dollars of agricultural and forest products. In 1921, Edward Forbush wrote that "forest and agricultural pests were reduced by 28% by birds resulting in savings of $444,000,000 in crop and timber losses." The value of birds in current dollars is beyond our imagination. Their value is not just in their actual consumption of insect pests, but also in their role in keeping future outbreaks to a minimum.

Further Reading:

Dickson, J.G. et. al., eds. *The Role of Insectivorous Birds in Forest cosystems*. Academic Press. New York. 1979. 381 pp.

Holling, C.S. *Temperate Forest Insect Outbreaks, Tropical Deforestation and Migratory Birds*. Mem. Entomol. Soc. Canada. 146:21-32. 1988

Morrison, M.L. et. al., eds. *Avian Foraging: Theory, Methodology, and Applications*. Studies in Avian Biology No. 13. Cooper Ornithological Society. Allen Press, Inc. Lawrence, KS. 1990. 514 pp.

Pschorn-Walker, H. *Biological Control of Insects*. Ann. Rev. Entomol. 22:1-22. 1977.

Takekawa, J.Y. et. al. *Biological Control of Forest Insect Outbreaks*: The Use of Avian Predators. in 47th N.A. Wildlife Conference. pp. 393-408.

Name: _____ Period: _____ Date: _____

How Birds Keep Our World Safe
Worksheet

1. What ecosystem service is discussed in this article?

2. How can a severe outbreak of insects affect a forest ecosystem?

3. What are some of the threats that insects face?

4. Name one adaptation used by insects to "dodge" death.

5. Birds are highly skilled and highly adapted insect-eaters. Give one example from the reading of this.

6. Give at least two examples of what we can do to help the birds that help us.

    •

    •

Name: _____ Period: _____ Date: _____

How Birds Keep Our World Safe
Worksheet Answer Key

1. What ecosystem service is discussed in this article?

    Pest control. Birds eat insects and reduce outbreaks that could destroy forests or crops.

2. How can a severe outbreak of insects affect a forest ecosystem?

    Insects can kill trees and destroy forests by eating leaves and attacking the bark.

3. What are some of the threats that insects face?

    Bad weather, disease, parasites, habitat destruction, insecticides, and predation.

4. Name one adaptation used by insects to "dodge" death.

    Some make themselves poisonous or produce spines or camouflage. Others have adapted ways of hiding themselves or evidence of where they have been feeding, such as feeding in places or at times of day that make them hard to detect.

5. Birds are highly skilled and highly adapted insect-eaters. Give one example from the reading of this.

    Possibilities from the article include:

    - *Selecting strong and healthy insects and avoiding those with parasites*
    - *Moving into areas with high insect population densities*
    - *Breeding during the peak period for insect populations, and possibly increasing the number of offspring when insect food availability is particularly high*
    - *Altering diet to feed almost exclusively on a particular insect species during an outbreak*
    - *Adapting foraging behavior in response to an outbreak, for example moving into the treetops when insects are particularly abundant there even though that bird species typically feeds at the ground level*

6. Give at least two examples of what we can do to help the birds that help us.

    Possibilities from the article include:

    - *Providing next boxes for cavity-nesting birds*
    - *Refraining from chopping down dead trees, which provide needed habitat for woodpeckers and other cavity-nesting birds*
    - *Integrated Pest Management (IPM), including providing habitat that supports birds and helps them to control insect pests*

WHY BIRDS MATTER

The Benefits of Birds to Humans and Nature

IN ONE HOUR

IN ONE DAY

= 60 INSECTS

The **BARN SWALLOW** can eat up to 60 insects per hour, and up to 850 in a day!

INSECT CONTROL

On its wintering grounds, the **YELLOW WARBLER** helps control harmful insects that attack coffee plantations.

Hummingbirds can visit more than 1,000 flowers in a day.

When a hummingbird feeds on flower nectar, pollen brushes off onto the bill and face. As the hummingbird moves from flower to flower, this pollen is transferred.

POLLINATION

BROAD-TAILED HUMMINGBIRD

Hummingbirds pollinate many flowering plants, including honeysuckle, columbine, fuchsia, penstemon, and aloe.

RUBY-THROATED HUMMINGBIRD

ARTISTIC INSPIRATION

The song of the European Starling inspired the 3rd movement of Mozart's Piano Concerto in G.

The expansive wings, dramatic courtship "dance", and beautiful white feathers and plumes of the **GREAT EGRET** inspire paintings, dances, poetry, and stories.

INDICATORS OF ENVIRONMENTAL CHANGE

AMERICAN PEREGRINE FALCON POPULATION

NUMBER OF BREEDING PAIRS

3000

Peregrine Listed Endangered 1970

Downlisted To Threatened 1984

Delisted 1999

2000

1000

0

1964 1972* 1980 1988 1996 2004

*DDT banned in the United States in 1972

The drastic decline of the **PEREGRINE FALCON** alerted us to the dangers of the pesticide DDT.

NATURE'S GARBAGE DISPOSAL

Finds carcasses using a keen sense of smell.

Featherless head stays cleaner when feeding on carrion.

Removes large amounts of decaying meat and kills bacteria and diseases in its highly acidic stomach.

The **TURKEY VULTURE** is a carrion specialist with unique adaptations to its lifestyle.

SMALL MAMMAL CONTROL

85%

A Red-tailed Hawk's diet consists of up to 85% rodents.

The other 15% consists of reptiles and birds.

15%

The **RED-TAILED HAWK** helps control rodent populations.

SEED DISPERSAL

The **AMERICAN ROBIN** plays an important role in dispersing seeds of shrubs and trees.

Robins eat a large variety of fruits and berries.

The seeds of these fruits pass through their digestive system and are deposited far from the original plant.

USGS

Online interactive version: http://www.birdday.org/images/stories/2014IMBD/infographicweb.swf

Created by Environment for the Americas for International Migratory Bird Day.

2 investigation

Exploring Habitat Needs of Nesting Birds

Investigation 2 Photo Credits

Kelly Colgan Azar. Belted Kingfisher. (https://flic.kr/p/8oAmcS)

Kelly Colgan Azar. Wood Thrush. (https://flic.kr/p/8aFWp6)

Kelly Colgan Azar. Red□Bellied Woodpecker. (https://flic.kr/p/dN8pDp)

Investigation 2
Exploring Habitat Needs of Nesting Birds

In exploring where various bird species nest in New York State, students discover that some species are found practically everywhere while others have limited distribution due to their more specialized habitat requirements. This Investigation is designed for use by classes in any state but focuses on nesting locations in New York because of the availability of detailed Breeding Bird Atlas Data and great variety of habitats ranging from boreal forests to marine and fresh water coastlines. The Investigation could be adapted to focus on any state for which similar data are available (see the USGS Breeding Bird Atlas Explorer http://www.pwrc.usgs.gov/bba/).

Essential Questions

- How do environmental factors influence where birds live and breed?

- How are species adapted to specific environmental conditions?

- How do environmental factors and species-specific adaptations interact to determine where birds live and breed?

Science Concepts and Topics

Some bird species are *generalists* while others are *specialists*. Generalist species are adapted to living under a wide range of environmental conditions. When it comes to nesting, this can be seen in comparing two species of chickadees. Black-capped Chickadee is a generalist species, nesting throughout the state. In contrast, Boreal Chickadee is a specialist species with a narrower ecological niche. It is one of the few bird species living exclusively in the northern boreal forest biome, which stretches across much of Canada and the northernmost parts of the United States. In New York, the boreal forest biome and the Boreal Chickadee are found only in the Adirondack Mountain region.

Looking at differences such as these leads to comparisons of the *adaptations* of species to their habitats. Students learn, for example, that Chimney Swifts nest only in chimneys or on rock faces, while Eastern Meadowlarks require grassland habitats in which to nest and raise their young.

Standards Addressed in This Investigation

Next Generation Science Standards – Middle School

- Analyze and interpret data to provide evidence for the effects of resource availability on organisms and populations of organisms in an ecosystem. (MS-LS2-1)

- Construct an explanation that predicts patterns of interactions among organisms across multiple ecosystems. (MS-LS2-2)

Next Generation Science Standard – Scientific Practices

- Asking questions and defining problems

- Planning and carrying out investigations

- Analyzing and interpreting data

- Constructing explanations and designing solutions

- Engaging in argument from evidence
- Obtaining, evaluating, and communicating information

Common Core (9-10ᵗʰ grade)

Reading Science & Technical Subjects

Key Ideas and Details

- Cite specific textual evidence to support analysis of science and technical texts, attending to the precise details of explanations or descriptions. (RST.9-10.1)
- Follow precisely a complex multistep procedure when carrying out experiments, taking measurements, or performing technical tasks, attending to special cases or exceptions defined in the text. (RST.9-10.3)

Craft and Structure

- Determine the meaning of symbols, key terms, and other domain-specific words and phrases as they are used in a specific scientific or technical context relevant to grades 9-10 texts and topics. (RST.9-10.4)

Integration of Knowledge and Ideas

- Compare and contrast findings presented in a text to those from other sources (including their own experiments), noting when the findings support or contradict previous explanations or accounts. (RST.9-10.9)

Range of Reading and Level of Text Complexity

- By the end of grade 10, read and comprehend science/technical texts in the grades 9-10 text complexity band independently and proficiently. (RST.9-10.10)

Writing

Research to Build and Present Knowledge

- Conduct short as well as more sustained research projects to answer a question (including a self-generated question) or solve a problem; narrow or broaden the inquiry when appropriate; synthesize multiple sources on the subject, demonstrating understanding of the subject under investigation. (W.9-10.7)
- Gather relevant information from multiple authoritative print and digital sources, using advanced searches effectively; assess the usefulness of each source in answering the research question; integrate information into the text selectively to maintain the flow of ideas, avoiding plagiarism and following a standard format for citation. (W.9-10.8)
- Draw evidence from literary or informational texts to support analysis, reflection, and research. (W.9-10.9)

Speaking and Listening

Comprehension and Collaboration

- Initiate and participate effectively in a range of collaborative discussions (one-on-one, in groups, and teacher-led) with diverse partners on grades 9-10 topics, texts, and issues, building on others' ideas and expressing their own clearly and persuasively. (SL.9-10.1)
- Integrate multiple sources of information presented in diverse media or formats (e.g., visually, quantitatively, orally) evaluating the credibility and accuracy of each source. (SL.9-10.2)

Exploring Habitat Needs of Nesting Birds
Before You Start...

Time

75-90 minutes, depending on what you ask of students in terms of the readings in this Investigation.

Preparation

- Make copies of the Worksheets.
- Access the *"Birds 2 Map"* at: http://crossingboundaries.org/bwb.php
- Follow these steps only if you will be using the Interactive PDF instead of ArcGIS Online:
  - Install Acrobat Reader from www.get.adobe.com/reader on student computers.
  - Download the *"Birds 2 Map"* interactive PDF from http://crossingboundaries.org/bwb.php.
  - Make this interactive PDF accessible on student computers and test the functionality of the layers.

Technology

A-V Equipment

- LCD Projector
- Screen or Whiteboard

Computers

- Web access to ArcGIS Online

If using Interactive PDF option:

- Acrobat Reader
- *"Birds 2 Map"* interactive PDF

Information and Communication Technology

- ArcGIS Online

Low-Tech Options

- Use the interactive PDF map
- If student computers are unavailable, print and/or photocopy the layers of the interactive PDF for use as transparent overlays or handouts.

Materials

- *"Birds 2 Map"* (online or interactive PDF)
- Web Map Help Sheet

Worksheets

- Using the Bird Atlas to Explore Bird Distribution
- Exploring Species-Specific Nesting Requirements

Exploring Habitat Needs of Nesting Birds
The Investigation

Investigation Overview

1. Introduce map layers and interpretation of spatial data.

2. Explore relationships between nesting distribution and environmental features.

3. Investigate adaptations and habitat needs of selected bird species.

4. Present and discuss species-specific relationships between adaptations and habitat needs.

Learning Objectives

Students will be able to:

- Interpret maps to make inferences about environmental features
- Analyze spatial data about bird distributions in relation to environmental features
- Explain nesting distribution patterns according to species-specific adaptations and habitat requirements.

Key Concepts

- Habitat
- Adaptation
- Generalist species
- Specialist species

Technology Overview

- Students use *NYS Breeding Bird Atlas* data and web resources to learn about the relationship between habitats and nesting bird distribution patterns.

Conducting the Investigation

1. Introduce map layers and interpretation of spatial data related to bird habitats.

    a. If the students have not yet used ArcGIS Online (or an Interactive PDF if you are using that option), have them open the **New York Breeding Bird Atlas Map** and practice turning layers on and off.

    b. Together as a class, work through the first four steps in the **Using the Bird Atlas to Explore Bird Distribution Worksheet**.

    c. After the students have seen the widespread distribution of Black-Capped Chickadees throughout New York State (step 4), ask if they think all bird species follow this same pattern or whether they can think of any that might be more selective in the types of habitat in which they build their nests.

2. Explore relationships between nesting distribution and environmental features.

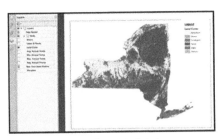

New York Breeding Bird Atlas

a. Together as a class, proceed to Step 5 in the worksheet and explore possible reasons for the patchy distribution of Red-Bellied Woodpeckers.

b. Discuss findings with regard to elevation and land cover, and have students answer the two questions in Step 6 about how these variables relate to nesting sites of this species in New York.

c. Individually or in small groups, have students complete steps 7 and 8 in the worksheet.

d. As a class, discuss how they categorized the various species and what questions they may have had while trying to relate environmental characteristics to distribution of nesting sites for each species.

3. Investigate adaptations and habitat needs of selected bird species.

a. Using Part A of the **Exploring Species-Specific Nesting Requirements Worksheet**, have each student use the map to select one bird species and make predictions about its nesting habitat requirements.

b. Using Part B of the worksheet, ask students to use the *All About Birds* website or other information sources to address questions related to habitat needs for their selected species.

4. Present and discuss species-specific relationships between adaptations and habitat needs.

a. In order to frame class discussion of the students' findings from the worksheet, create a two column T-chart on the board with one side for generalist species and the other side for specialist species.

b. Ask students to place the species they investigated into one of these two columns based on the extent to which its nesting sites are limited to specific types of habitat. Species that nobody has investigated in detail can be placed on the chart based on students' answers to Question 8 in the first worksheet (Using the **Bird Atlas to Explore Bird Distribution Worksheet)**.

c. Discuss the students' responses.

Teacher Notes about Selected Bird Species

Species	Nesting Habitat	Nesting Range in New York	Range beyond New York	Notes
Barred Owl http://www.allaboutbirds.org/ guide/Barred_Owl/id	Tree cavities in wooded areas and swamps, favoring large blocks of forest	Widespread, but clustered in mountains and areas with dense woods	Year-round throughout eastern US, across Canada, and in western Mexico	Barred Owls are fairly numerous and their range has been expanding westward through Canada over the past century. Because they need large, dead trees for nest sites, the Barred Owl is often used as an indicator species for managing old forests.
Belted Kingfisher http://www.allaboutbirds.org/ guide/Belted_Kingfisher/id	Burrows dug into banks along or up to 2 km from streams, lakes, and calm marine waters	Widespread	Year-round in much of the US, + summer ranges in Canada and winter in Central America	Belted Kingfishers perch along streams, lakes, and estuaries, searching for small fish. They hunt by plunging from a perch or by hovering over the water before diving after a fish they've spotted. Kingfishers spend winters in areas where the water doesn't freeze so they can continue to catch aquatic prey.
Black-backed Woodpecker http://www.allaboutbirds.org/ guide/Black-backed_Woodpecker/id	Holes in tree trunks in coniferous northern boreal forests, especially in areas with burned trees	Only in the Adirondack Mountains	Year-round in Canada, Pacific NW, and northernmost parts of NY and New England	An uncommon woodpecker of the northern coniferous forests, the Black-backed Woodpecker prefers burned-over sites. It moves from place to place, following outbreaks of wood-boring beetles in recently burned habitats.

Teacher Notes about Selected Bird Species

Species	Nesting Habitat	Nesting Range in New York	Range beyond New York	Notes
Black-capped Chickadee http://www.allaboutbirds.org/guide/Black-capped_Chickadee/id	Small tree cavities or nest boxes in forests, fields, parks, and suburban areas	Widespread	Year-round in Canada and northern half of the US	Black-capped Chickadees are commonly seen at bird feeders and are found in any habitat that has trees or woody shrubs, from forests and woodlots to residential neighborhoods and parks. Sometimes found in weedy fields or cattail marshes.
Boreal Chickadee http://www.allaboutbirds.org/guide/Boreal_Chickadee/id	Spruce-fir forests	Only in the Adirondack Mountains	Year-round in Canada and northernmost parts of the contiguous US	The Boreal Chickadee is one of the few birds living completely within the northern boreal forest biome in Canada and northern US.
Carolina Wren http://www.allaboutbirds.org/guide/Carolina_Wren/id	Brushy cover in many settings, including suburbs	Scattered across the state except in the mountains	Year-round in eastern US and the Yucatan	The Carolina Wren is sensitive to cold weather, and NY is at the northeastern limit of its distribution.
Chimney Swift http://www.allaboutbirds.org/guide/Chimney_Swift/id	Originally nested in hollow trees and perhaps caves but now primarily in chimneys	Scattered throughout the NY State	Migrates between eastern US and NW South America	Populations are decreasing because new chimneys are less suitable as nest sites. The Chimney Swift uses glue-like saliva from a gland under its tongue to cement its nest to the chimney wall or rock face. The young outgrow the nest after about two weeks and cling to the nearby wall, in many cases even before their eyes are open.

Teacher Notes about Selected Bird Species

Species	Nesting Habitat	Nesting Range in New York	Range beyond New York	Notes
Common Raven http://www.allaboutbirds.org/guide/Common_Raven/id	On cliffs, in trees, and on structures such as power-line towers, telephone poles, billboards, and bridges.	Widespread except in lowlands.	Year-round across the northern hemisphere and into Central America	Common Ravens live in open and forest habitats across western and northern North America. This includes deciduous and evergreen forests up to treeline, as well as high desert, sea coast, sagebrush, tundra, and grasslands.
Eastern Meadowlark http://www.allaboutbirds.org/guide/Eastern_Meadowlark/id	Nest on the ground in meadows or grassy pastures, typically well concealed by dense vegetation	Widespread in open areas. Rare on Long Island & mountains.	Migratory in northern states in the eastern US and year-round in more southern states down to northern parts of South America	Eastern Meadowlarks are a declining species because grassland habitat is disappearing and small family farms with pastureland and grassy fields are being replaced by larger, row-cropping agricultural operations or by development. Early mowing, overgrazing by livestock, and pesticides use also harm meadowlarks.
Eastern Screech-Owl http://www.allaboutbirds.org/guide/Eastern_Screech-Owl/id	Tree cavities or nest boxes in fragmented forests, farmland, suburban landscapes, and city parks	Predominantly in fragmented forests	Year-round in eastern US. NY is at the northeastern limit of the range.	Eastern Screech-Owls do not dig nest cavities so depend on tree holes opened or enlarged by woodpeckers, fungus, rot, or squirrels. They build no nest, simply making a body-shaped depression and laying eggs on whatever debris is at the bottom of the nesting cavity.
Eastern Whip-poor-will http://www.allaboutbirds.org/guide/whip-poor-will/id	In forests close to open areas, avoiding large tracts of uninterrupted forest with dense canopy	Declining, found in barrens & fragmented forests	In summer across much of the eastern US. Migrate to Mexico and Central America for the winter.	These common birds are on the decline in parts of their range as open forests are converted to suburbs or agriculture. They lay their eggs to hatch on average 10 days before a full moon. When the moon is near full, the adults can forage the entire night and capture large quantities of insects to feed to their nestlings.

Teacher Notes about Selected Bird Species

Species	Nesting Habitat	Nesting Range in New York	Range beyond New York	Notes
Marsh Wren http://www.allaboutbirds.org/guide/Marsh_Wren/id	Nests of grasses and sedges are lashed to vegetation in fresh and salt water marshes	Along coasts and rivers	Year-round in several western states and migratory across most of the US	Nests only in wetland habitats.
Mourning Dove http://www.allaboutbirds.org/guide/Mourning_Dove/id	Open habitats, including farm fields, forest edges, cities, and suburbs	Widespread except where forest cover is continuous	Year-round across most of the US and migratory in the upper Midwest and Central America	One of the most widespread and abundant birds in North America, Mourning Doves can be seen nearly anywhere except in deep woods. This species was relatively rare in New York before European settlers converted the extensive forests to farmland.
Olive-sided Flycatcher http://www.allaboutbirds.org/guide/Olive-sided_Flycatcher/id	Edges and openings in coniferous or mixed woods, along ponds and rivers, and in burned-over forests	Limited to mountainous regions in NY	Migratory from western US and northern Canada to wintering grounds in Central and South America	New York is at the southeastern border of the breeding range for this species. Populations are declining. A likely cause is deforestation in its winter range. Another cause may be decline in bee populations, a favored food.

Teacher Notes about Selected Bird Species

Species	Nesting Habitat	Nesting Range in New York	Range beyond New York	Notes
Peregrine Falcon http://www.allaboutbirds.org/guide/Peregrine_Falcon/id	Rocky cliffs and also buildings, bridges, towers, and nesting platforms	In NYC, western end of Long Island, and along the Hudson River.	Year-round in western US. The Peregrine has one of the longest migrations of any North American bird, stretching from the Arctic to South America	One of the world's most widely distributed bird species, present on all continents except Antarctica. DDT led to population crash, but captive breeding has helped in recovery and the species was removed from the federal endangered species list in 1999. About half of NY's peregrine nests are on cliffs and half on buildings or bridges.
Red-bellied Woodpecker http://www.allaboutbirds.org/guide/Red-bellied_Woodpecker/id	Holes in tree trunks in mixed deciduous-coniferous forests and suburbs	Absent in NY's mountains, primarily found in areas with a mean July temp > 70°F	Year-round in eastern US	NY is at the northeastern edge of the range for the Red-bellied Woodpecker. They often appear at feeders and are common across most of the forests, woodlands, and wooded suburbs of the eastern United States.
Rock Pigeon http://www.allaboutbirds.org/guide/Rock_Pigeon/id	On urban window ledges and under bridges, in barns and grain towers, and on natural cliffs	Widespread except where forest cover is continuous	Introduced worldwide and a year-round resident in South & Central America, US, & southern Canada	Domesticated over 5000 years ago, Rock Pigeons were introduced to North America from Europe in the 1600s as a source of food. This is the common pigeon now widespread and abundant, especially in cities.

Teacher Notes about Selected Bird Species

Species	Nesting Habitat	Nesting Range in New York	Range beyond New York	Notes
Ruby-throated hummingbird http://www.allaboutbirds.org/guide/Ruby-throated_Hummingbird/id	On branches or other surfaces in forests, gardens, yards, and orchards	Widespread except in New York City and western Long Island	Migratory, summering in eastern US or southern Canada and wintering in Central America	The Ruby-throated Hummingbird is eastern North America's sole breeding hummingbird. Each fall they migrate to Central America, with many crossing the Gulf of Mexico in a single flight.
Wood Duck http://www.allaboutbirds.org/guide/Wood_Duck/id	Tree cavities and nest boxes in bottomland forests, swamps, beaver flows, and slow streams	Scattered throughout NY State	Year-round in much of eastern US and some western states. Migratory in northern New York and some other states to winter in Central America	Close to extinction in the early 1900s, Wood Ducks recovered due to reversion of farmlands to forest, legal protection from over-hunting, increasing beaver populations creating more ponds, and ambitious nest box programs. They cannot make their own cavities but can nest high in trees up to 2 km from water.
Wood Thrush http://www.allaboutbirds.org/guide/Wood_Thrush/id	In lower branches of deciduous forests that have high canopy, understory, and a leaf litter layer	Widespread except in the Adirondack Mountains	Migratory between eastern US and lowland tropical forests in Central America	Wood Thrush populations are declining in most regions across its range, with steep declines in Atlantic Coast and New England states where Wood Thrushes are most common. One reason may be habitat fragmentation in both breeding and wintering grounds. Another is acid rain, which can leach calcium from the soil and rob the birds of vital, calcium-rich invertebrate prey.

Sources: *All About Birds* (http://www.allaboutbirds.org/) and McGowan, K.J. and Corwin, K. (Eds.) 2008. *The Second Atlas of Breeding Birds in New York State*. Ithaca, NY: Cornell University Press.

Web Map Help Sheet
NYS Breeding Bird Atlas

A guide to navigation and use of the web map for this investigation.

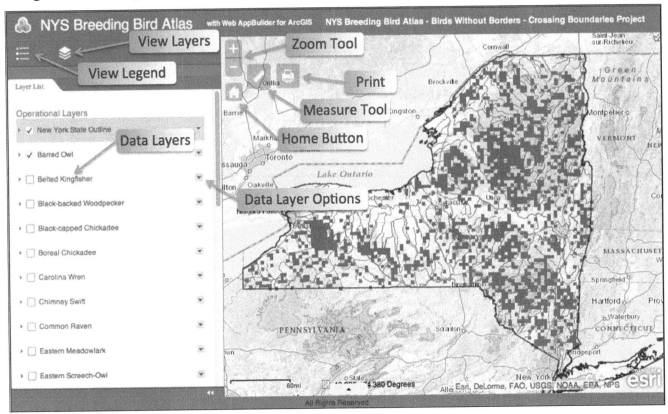

The web maps in this unit provide access to real data about species distributions and environmental conditions, landforms, and other geographical features. Each map includes a series of "data layers." You can view one layer at a time or turn on several simultaneously to explore how they relate.

When you first open a web map, the list of available data layers may be hidden. To make it visible, simply click on the View Layers icon. When a layer is turned on, its Legend becomes active. You can toggle back and forth between viewing the layer list and legends simply by clicking on the View Layers and View Legend icons.

The Data Layer Options symbol brings up a menu offering options including "Transparency." Try moving the Transparency slider and explore how that helps you compare two or more data layers.

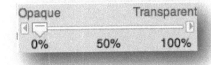

Icons in the top left corner of the map show available tools. All maps include tools for zooming in and out, measuring features, and printing your map. The Home button will bring the map back to the original perspective seen when it was first opened.

The best way to learn how to use one of these maps is to play with it. You can't break it!

Using the Breeding Bird Atlas to Explore Bird Distribution
Worksheet

1. Turn on the "Land Cover" layer.

    On the outline map below, draw a circle to show the location of the largest area of forest in the state.

    Make a rectangle to show the location of the largest area of farmland.

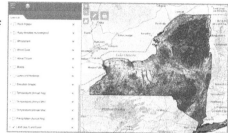

2. Turn on and off the various temperature layers and put Xs on your map to show where the coldest temperatures occur.

3. Using a map of the United States, take a look at New York State's borders. Label on your map these shorelines:

   - Lake Erie
   - Lake Ontario
   - Atlantic Ocean

4. Check the box next to "Black-Capped Chickadee" to turn on this layer. Each square that is colored brown indicates an area in which chickadees were found nesting.

   a. Where do chickadees breed in New York State?

   b. Would you call this species a *generalist* or a *specialist*? Why?

5. Turn off the "Black-capped Chickadee" layer and turn on the "Red-bellied Woodpecker" layer. What do you notice about the woodpecker's distribution?

6. Use the environmental data to explore which factors may be determining where Red-bellied Woodpeckers nest in New York. First, turn off the woodpecker layer and look only at "Elevation." Toggle the woodpecker layer on and off several times and see what you can conclude about elevations at which nests are found for this species. Now turn off "Elevation" and try the same steps with "Land Cover." What can you conclude about Red-bellied Woodpecker nesting habitat in terms of the following factors?

   a. Elevation:

   b. Land Cover:

7. Compared with the Black-capped Chickadee, is the Red-bellied Woodpecker more of a generalist or a specialist species?

   On what evidence did you base this conclusion?

8. Now explore nesting distribution of the other bird species. Make sure to turn on only one species at a time. Note how the distribution of each species relates to the landscape features you explored earlier. See if you can find at least one species for each of these categories:

   a. Nest almost uniformly across the entire state:

   b. Nest mostly at high elevations:

c. Nest mostly at low elevations:

d. Nest mostly along waterways or coastlines:

e. Not seen nesting in urban areas:

f. Nest in urban areas (and also in other types of habitat):

g. Nest primarily in forested areas:

h. Nest primarily in non-forested areas:

i. Nest in areas with warmer temperatures:

Using the Breeding Bird Atlas to Explore Bird Distribution
Worksheet Answer Key

1. Turn on the "Land Cover" layer.

   On the outline map below, draw a circle to show the location of the largest area of forest in the state.

   Make a rectangle to show the location of the largest area of farmland.

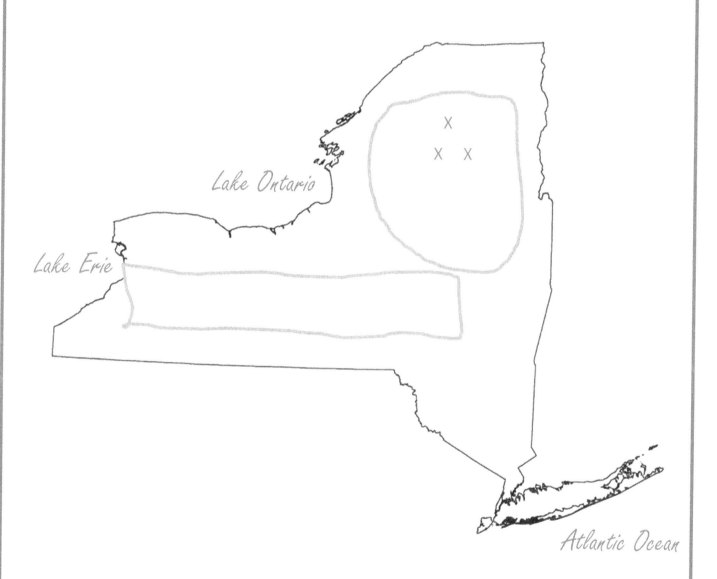

2. Turn on and off the various temperature layers and put Xs on your map to show where the coldest temperatures occur.

3. Using a map of the United States, take a look at New York State's borders. Label on your map these shorelines

   - Lake Erie
   - Lake Ontario
   - Atlantic Ocean

4. Now open the "Birds" folder (by clicking on the triangle) and then turn on the "Black-capped Chickadee" layer. Each square that is colored brown indicates an area in which chickadees were found nesting.

   a. Where do chickadees breed in New York State?

   Black-capped Chickadees were found nesting almost everywhere in NYS.

   b. Would you call this species a **generalist** or a **specialist**? Why?

   Generalist, because they nest in a wide range of habitats rather than being limited to a specific habitat type.

5. Turn off the "Black-capped Chickadee" layer and turn on the "Red-bellied Woodpecker" layer. What do you notice about the woodpecker's distribution?

   There are large parts of the state in which this species was not found to be nesting.

6. Use the environmental data to explore which factors may be determining where Red-bellied Woodpeckers nest in New York. First, turn off the woodpecker layer and look only at "Elevation." Toggle the woodpecker layer on and off several times and see what you can conclude about elevations at which nests are found for this species. Now turn off "Elevation" and try the same steps with "Land Cover." What can you conclude about Red-bellied Woodpecker nesting habitat in terms of the following factors?

    a. Elevation:

    Red-bellied Woodpeckers nest at the lower elevations and avoid the mountainous parts of the state.

    b. Land Cover:

    Most of the agricultural areas in New York are used as nest sites by this species. Developed areas also used, and also some of the forest regions.

7. Compared with the Black-capped Chickadee, is the Red-bellied Woodpecker more of a generalist or a specialist species?

    The Red-bellied Woodpecker is more of a specialist species, shown in the more restricted range of habitats in which it is found nesting in New York State.

    On what evidence did you base this conclusion?

    Answers vary

8. Now explore nesting distribution of the other bird species. Make sure to turn on only one species at a time. Note how the distribution of each species relates to the landscape features you explored earlier. See if you can find at least one species for each of these categories:

    a. Nest almost uniformly across the entire state:

    Black-capped Chickadee

    Ruby-throated Hummingbird

    b. Nest mostly at high elevations:

    Black-backed Woodpecker

    Boreal Chickadee

    Olive-sided Flycatcher

c. Nest mostly at low elevations:

Carolina Wren

Eastern Meadowlark

Eastern Screech-Owl

Mourning Dove

Red-bellied Woodpecker

Rock Pigeon

Wood Thrush

d. Nest mostly along waterways or coastlines:

Marsh Wren

Peregrine Falcon

(Belted Kingfisher and Wood Duck also nest near water, but these species are more broadly distributed and their dependence on water is harder to figure out from the maps.)

e. Not seen nesting in urban areas:

Barred Owl

Boreal Chickadee

Common Raven

f. Nest in urban areas (and also in other types of habitat):

Chimney Swift

g. Nest primarily in forested areas:

Barred Owl

Common Raven

h. Nest primarily in non-forested areas:

Eastern Screech-Owl

i. Nest in areas with warmer temperatures:

Carolina Wren

Name: _____ Period: _____ Date: _____

Exploring Species-Specific Nesting Requirements
Worksheet

Select one bird species to investigate. What species did you select? _____

Part A: Making Predictions about Your Species

Answer these questions based on your work with the map:

1. In what type of area does this species nest?

2. In what type of area is your species NOT found during nesting season?

3. Write a sentence or two explaining why you think this species might be distributed in this way in New York State.

4. What type of habitat do you think your species needs during the nesting season?

Part B: Investigate the habitat requirements for your species

Answer these questions using information from *All About Birds* (http://allaboutbirds.org/) and other sources:

1. What does your species eat?

2. In what type of habitat does it live during nesting season? (forest, grassland, sandy beach, etc.)

3. Where does your species nest? (on the ground, in a tree, etc.)

4. Does it have any special behaviors or adaptations related to its habitat?

Part C. Compare your inferences with your findings:

Go back and read the inferences you made in Part A, steps 1 and 2. How does your prediction compare with what you learned from this investigation into habitat requirements of your species during breeding season?

Extension: Thinking About Scientific Data Collection

The **New York Breeding Bird Atlas Map** was created using data collected by more than 1,200 volunteers between 2000 and 2005. The state was divided into 5,332 blocks, each measuring 5x5 km. Volunteers visited various habitats within their assigned blocks and recorded evidence of breeding for each bird species that they found.

Why was it important for all of the volunteer scientists to follow the same instructions in collecting their field data?

3 investigation

Determining Annual Life Cycles of Local Birds

American Goldfinch
Spinus tristis

Investigation 3 Photo Credits

Kevin J. McGowan. Northern Parula. Cornell Lab of Ornithology.

Phil Kahler. Northern Pintail. (https://flic.kr/p/dNVHZE)

Andy Johnson. Hudsonian Godwit. (andyjohnsonphoto.com)

Investigation 3
Determining Annual Life Cycles of Local Birds

Students may be surprised to learn that some bird species live in their region year-round while others migrate seasonally. Using data visualization tools provided by the eBird citizen science project, they discover which species fall into these two categories in their state or county.

Essential Questions

- Why do some birds migrate while others do not?
- How does this relate to each species' habitat requirements during nesting season and throughout the year?

Science Concepts and Topics

Do all birds *migrate*? Of the approximately 10,000 bird species worldwide, over 600 breed in North America. Some of these stay in one region throughout their entire life cycle while others migrate seasonally. Of the migrants, some fly to distant lands while others make journeys as short as up and down in elevation on a single mountain.

What *habitat requirements* cause some species to exert all the energy needed to leave the tropics and migrate north for breeding season? The primary reason for undertaking these long and risky journeys is to find ample high-protein food to feed their young. Canada and the U.S. provide a wealth of insects along with space in which to spread out and longer daylight hours in which to search for food. Species that stay in North America year-round are those with *adaptations* enabling them to find seeds, berries, or other food during our winter months.

Data used in this Investigation were collected through eBird, a citizen science project in which anyone anywhere in the world can submit data about birds they have observed. Collectively, these observations by citizen scientists are creating a massive database of use to professional scientists, birders, students, and anyone interested in exploring distribution of almost any of the world's over 10,000 species of birds.

Standards Addressed in This Investigation

Next Generation Science Standards – Middle School

- Construct an explanation that predicts patterns of interactions among organisms across multiple ecosystems. (MS-LS2-2)

Next Generation Science Standards – High School

- HS-LS2-8. Evaluate the evidence for the role of group behavior on individual and species' chances to survive and reproduce.

Next Generation Science Standards – Scientific Practices

- Asking questions and defining problems
- Planning and carrying out investigations
- Analyzing and interpreting data
- Constructing explanations and designing solutions
- Engaging in argument from evidence

- Obtaining, evaluating, and communicating information

Common Core (9-10th grade)

Reading Science & Technical Subjects

Key Ideas and Details

- Cite specific textual evidence to support analysis of science and technical texts, attending to the precise details of explanations or descriptions. (RST.9-10.1)

- Determine the central ideas or conclusions of a text; trace the text's explanation or depiction of a complex process, phenomenon, or concept; provide an accurate summary of the text. (RST.9-10.2)

- Follow precisely a complex multistep procedure when carrying out experiments, taking measurements, or performing technical tasks, attending to special cases or exceptions defined in the text. (RST.9-10.3)

Craft and Structure

- Determine the meaning of symbols, key terms, and other domain-specific words and phrases as they are used in a specific scientific or technical context relevant to grades 9-10 texts and topics. (RST.9-10.4)

Integration of Knowledge and Ideas

- Translate quantitative or technical information expressed in words in a text into visual form (e.g., a table or chart) and translate information expressed visually or mathematically (e.g., in an equation) into words. (RST.9-10.7)

- Compare and contrast findings presented in a text to those from other sources (including their own experiments), noting when the findings support or contradict previous explanations or accounts. (RST.9-10.9)

Range of Reading and Level of Text Complexity

- By the end of grade 10, read and comprehend science/technical texts in the grades 9-10 text complexity band independently and proficiently. (RST.9-10.10)

Writing

Research to Build and Present Knowledge

- Conduct short as well as more sustained research projects to answer a question (including a self-generated question) or solve a problem; narrow or broaden the inquiry when appropriate; synthesize multiple sources on the subject, demonstrating understanding of the subject under investigation. (W.9-10.7)

- Gather relevant information from multiple authoritative print and digital sources, using advanced searches effectively; assess the usefulness of each source in answering the research question; integrate information into the text selectively to maintain the flow of ideas, avoiding plagiarism and following a standard format for citation. (W.9-10.8)

- Draw evidence from literary or informational texts to support analysis, reflection, and research. (W.9-10.9)

Speaking *and* Listening

Comprehension and Collaboration

- Initiate and participate effectively in a range of collaborative discussions (one-on-one, in groups, and teacher-led) with diverse partners on grades 9-10 topics, texts, and issues, building on others' ideas and expressing their own clearly and persuasively. (SL.9-10.1)

- Integrate multiple sources of information presented in diverse media or formats (e.g., visually, quantitatively, orally) evaluating the credibility and accuracy of each source. (SL.9-10.2)

Determining Annual Life Cycles of Local Birds

Before You Start...

Time

Preparation: **15 minutes**

Instruction: **45 minutes**

Preparation

- Make copies of the Student Reading and Worksheet.

Technology

A-V Equipment

- LCD Projector
- Screen or Whiteboard

Computers

- Acrobat Reader
- Internet Connection

Information and Communication Technology

- eBird data visualization tools

Low-Tech Options

- Use printed copies of the eBird bar chart for your county and of range maps for selected species.

Materials

Student Reading

- Neotropical Migratory Birds

Worksheet

- Investigating Migration Using eBird Data

Determining Annual Life Cycles of Local Birds
The Investigation

Investigation Overview

1. Introduce the concept of migration.

2. Determine which bird species in your area migrate and which do not.

3. Investigate the habitat needs of an individual bird species.

4. Compile species information, discuss student responses, and synthesize into understandings about migration.

Learning Objectives

Students will be able to:

- Use eBird citizen science data and visualization tools to determine which bird species in their area migrate and which are year-round residents.

- Explain why long distance migration makes sense in the life cycle of some bird species.

Key Concepts

- Life Cycles
- Habitat
- Adaptation
- Migration

Technology Overview

- Students use eBird's data visualization tools and look up species-specific information on the *All About Birds* website.

Conducting the Investigation

1. Introduce the concept of migration.

   a. Ask students what birds they have seen in your area during the winter.

   Depends on where you live, but some common examples might include American Crow, Northern Cardinal, House Finch, Blue Jay, Black-capped Chickadee.

   b. Ask whether they have noticed any bird species in the spring and summer that don't seem to be around in winter months.

   Again depends on your location. You may want to use eBird to find the migratory species for your area in order to have a few common species in mind during this discussion (see Step 2 below).

   c. Compile two lists – one for birds seen in spring and summer and the other for birds seen in winter. You will refer back to these lists later in the Investigation.

   d. Ask students to read **Why Do Birds Migrate?** and then lead class discussion on the following questions:

- What is a neotropical migratory bird?

  A species that breeds in Canada or the U.S. and heads south to Mexico, Central America, South America or the Caribbean Islands for nonbreeding seasons.

- At what times of year do they migrate?

  In the spring they migrate north to lay eggs and raise young, and in the fall they head south to the tropics where they live during our winter months.

- Why do they migrate?

  Birds migrate to move to areas where resources are more abundant. The two primary resources are food and nesting locations. Some species can raise more young by migrating north for the breeding season because of the higher abundance of protein-rich insects and other food to feed their young. Other factors include longer daylight hours in which to search for food, greater area in which to spread out, and possibly fewer predators.

2. Determine which bird species in your area migrate and which do not.

   a. Project the eBird website (http://ebird.org), and demonstrate how to select the "Explore Data" tab and then the "Bar Charts" option. Select your state and county to create a bar chart showing what bird species have been reported there. Scroll down to the bottom of the page to see the key, and discuss with students how to interpret the chart.

   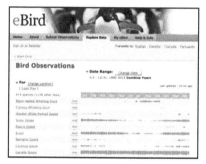

   http://ebird.org

   Each row represents the presence or absence of a single bird species. If the bars stretch across the entire year, that species is found in your county year-round. Shorter bars represent fewer sightings of that species than taller bars.

   b. Practice using the eBird bar charts to determine when various migratory bird species are found in your county. Ask students to complete Part A of the **Investigating Migration Using eBird Data Worksheet** to identify species found in their county:

   - year-round
   - during the summer breeding season
   - only in winter
   - during spring and fall migration periods

   c. Look for species in the lists that the class brainstormed in Step 1, and compare these with findings from the eBird bar charts. How well do they match?

   (Note: Some species may be seen there only during spring and fall as they move through on their way to other sites. Others may be present in the summer, or winter, but not during the rest of the year.)

3. Investigate the habitat needs of a specific selected species.

   a. Once students are familiar with how to interpret eBird's bar charts, have them complete Part B in the worksheet. They will choose representative species present in your county at designated times of year and then investigate the habitat needs of one of these species.

b. After completing Part B in the worksheet, they should also answer the reflection question in Part C.

4. Compile species information, discuss student responses, and synthesize into understandings about migration.

a. After students have completed the worksheet, reconvene to discuss their responses, especially to the final reflection question. Be sure to focus on the adaptations that make it possible for some bird species to stay in the same place year-round while others migrate between nesting and wintering grounds.

5. Possible extensions:

a. Have students explore the conservation status of their selected species and any threats to its abundance or distribution. This information is available under the "Life History" tab for each species in All About Birds (http://www.AllAboutBirds.org).

http://www.allaboutbirds.org

b. View Nate Senner's "In the Field" video and discuss why it is important for him to discover where Hudsonian Godwits fly on their long migratory journeys. Possible discussion questions include:

- What were your thoughts about Nate's research?

- Why is he investigating the migratory routes of this bird species?

- In what ways does his research help to determine places that should be protected?

- What other questions do you have about the habits and life histories of Hudsonian Godwits?

Nate Senner: *In the Field*

Neotropical Migratory Birds
Student Reading

What is a Neotropical migratory bird?

A Neotropical migratory bird is a bird that breeds in Canada and the United States during summer and then heads south to spend our winter months in Mexico, Central America, South America, or the Caribbean islands. These are places where it is spring or summer during our fall or winter seasons. Most Neotropical migratory birds are songbirds (such as warblers, thrushes, tanagers, and vireos). Others include shorebirds (such as sandpipers and plovers), some raptors (such as hawks, kites and vultures), and a few types of waterfowl (such as teal).

How far do they travel?

Migration distances vary greatly between species and between individual birds of the same species. Short migration journeys are made by birds that breed in the southern United States and fly south to Mexico or the West Indies. These trips can be as short as a few hundred kilometers. Examples of birds that make such relatively short migrations include Black-capped Vireos and Lucy's Warblers, and some Painted Buntings, Northern Parulas, and Gray Catbirds.

Some of the longest migrations are made by shorebirds that nest in the arctic tundra of northernmost Canada. Some of these species fly up to 16,000 kilometers twice each year between breeding sites in the north and non-breeding habitats in South America. Red Knots and White-rumped Sandpipers are examples of species that make these remarkable journeys. Other birds that fly great distances to nest in the U.S. or Canada include: Common Nighthawks, Swainson's Hawks, Red-eyed Vireos, Purple Martins, Barn and Cliff Swallows, Blackpoll, Cerulean and Connecticut Warblers, Scarlet Tanagers, and Bobolinks.

Why do they fly so far?

Because it's too far to walk. Seriously, the best explanation for why birds fly such great distances is to take advantage of seasonally abundant food and to avoid times or places where food and other resources are scarce.

You may have guessed that they migrate south to avoid the cold of our winter, but many bird species can tolerate cold temperatures as long as food is plentiful. Neotropical migratory birds need food such as flying insects, caterpillars, fruits and nectar to feed their young. These sources are super-abundant during our spring and summer seasons but not during our harsh winter months.

Ultimately, the reason why birds migrate is that they are likely to be able to raise more young by migrating than they would if they remained in the tropics. Reasons include the higher abundance of protein-rich insects and other food to feed their young. Their northern breeding areas also have longer daylight hours in which to search for food, more space in which to spread out, and possibly fewer predators.

This article was summarized from the Smithsonian Migratory Bird Center's fact sheet:

"Neotropical Migratory Bird Basics," which includes fascinating details about how far, how high, and how fast migratory birds can fly and how birds know when to migrate and where to go. The full articles is at http:// nationalzoo.si.edu/scbi/migratorybirds/fact_sheets/ default.cfm?id=9.

Name: _____ Period: _____ Date: _____

Investigating Migration Using eBird Data
Worksheet

On the eBird website (http://ebird.org), select the **Explore Data** tab and then the **Bar Charts** function. Select your state, click the button for "Counties in [your state]," and then select your county from the list provided.

Part A: Exploration

Use eBird's bar charts to identify bird species that fit into the following categories. Report your state and county below and then shade in the annual chart with green bars to represent the times of year in which that species has been seen in your county.

State: _____

County: _____

Category 1 - A species found in your county **year-round**

Species Name	Jan	Feb	Mar	Apr	May	Jun	Jul	Aug	Sep	Oct	Nov	Dec

Category 2 - A species that migrates to your county for the **summer breeding season**

Species Name	Jan	Feb	Mar	Apr	May	Jun	Jul	Aug	Sep	Oct	Nov	Dec

Category 3 - A species that is present in your county only in **winter**

Species Name	Jan	Feb	Mar	Apr	May	Jun	Jul	Aug	Sep	Oct	Nov	Dec

Category 4 - A species that passes through during **spring and fall migration periods**

Species Name	Jan	Feb	Mar	Apr	May	Jun	Jul	Aug	Sep	Oct	Nov	Dec

Part B: Species-Specific Research

Select a bird species from one of the four categories in Step 1. Investigate your selected species using All About Birds (http://www.allaboutbirds.org/), complete the worksheet below, and be prepared to discuss specific adaptations and habitat needs that may help to explain why it does or does not migrate.

Common name	
Scientific name	
Size, shape, and color (sketch, insert photo, or describe what this bird looks like)	
Migratory status	☐ Present year-round ☐ Present only in summer ☐ Present only in winter ☐ Present only in spring and fall
Range map (Shade in with colors coded by season, or insert appropriate re-sized screenshot) Color Season _____ _____ _____ _____ _____ _____ _____ _____	

Nesting habitat (such as forest, suburban yards, open fields, beaches)	
Food	
Behavior	
Conservation status	

Part C: Reflection: Answer one of the two questions below.

If your selected species stays in the same place year-round, what adaptations do you think make this possible?

If your species migrates, what adaptations and habitat needs explain this behavior?

4 investigation

Modeling Bird Population Trends

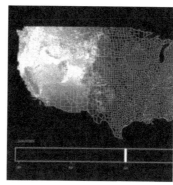

Investigation 4 Photo Credits

Kevin J. McGowan. Painted Bunting. Cornell Lab of Ornithology.

Kevin J. McGowan. Northern Cardinal. Cornell Lab of Ornithology.

Kevin J. McGowan. Scarlet Tanager. Cornell Lab of Ornithology.

Investigation 4
Modeling Bird Population Trends

In this Investigation, students use animated maps produced through modeling to view seasonal migration patterns of several related bird species, and they consider the advantages and limitations of modeling compared with simply using bird sighting data provided by citizen scientists. Investigation 5 builds on this theme by having students compare the model outputs with citizen science data representing sightings of the same four bird species.

This Investigation focuses on four bird species in the Cardinalidae family (Subphylum Vertebrata, Class Aves, Order Passeriformes). The Cardinalidae family includes cardinals, buntings, and some grosbeaks. Cardinalidae share the characteristic of thick, cone-shaped beaks used for feeding on seeds, fruit, and insects. They live in brushy or wooded habitats. The males typically are brightly colored while females have more muted colors or shades of brown. The four species highlighted here were selected because of differences in their migration patterns.

Essential Question

- What can modeling tell us about bird population dynamics?

Science Concepts and Topics

In *citizen science*, volunteers partner with professional scientists to collect or analyze data. In the eBird citizen science project, for example, birders around the globe contribute their sighting data to a central database that professional scientists use for research. Students and other members of the public can use the eBird database to keep track of their own sightings and to explore data entered by others.

Although the eBird database is vast and growing rapidly, gaps remain in areas or times of year in which birding is rare. *Modeling helps to fill data gaps*, combining the bird sighting data generated through citizen science with environmental data describing habitat conditions. *Modeling also makes it possible to predict* into the future, for example to investigate possible impacts of environmental change. Both eBird data and the model outputs are useful in investigating population dynamics of individual bird species, looking at which species migrate and where they are found at various times of year. This Investigation makes use of animated maps that vividly portray the results of modeling in which eBird citizen science data are combined with environmental variables to predict the timing and locations in which individual bird species are likely to be seen.

Standards Addressed in This Investigation

Next Generation Science Standards – Middle School

- Analyze and interpret data to provide evidence for the effects of resource availability on organisms and populations of organisms in an ecosystem. (MS-LS2-1)
- Construct an explanation that predicts patterns of interactions among organisms across multiple ecosystems. (MS-LS2-2)

Next Generation Science Standards Scientific Practices

- Asking questions and defining problems
- Developing and using models

- Planning and carrying out investigations
- Analyzing and interpreting data
- Constructing explanations and designing solutions
- Engaging in argument from evidence
- Obtaining, evaluating, and communicating information

Common Core (9-10th grade)

Reading Science & Technical Subjects

Key Ideas and Details

- Cite specific textual evidence to support analysis of science and technical texts, attending to the precise details of explanations or descriptions. (RST.9-10.1)
- Determine the central ideas or conclusions of a text; trace the text's explanation or depiction of a complex process, phenomenon, or concept; provide an accurate summary of the text. (RST.9-10.2)
- Follow precisely a complex multistep procedure when carrying out experiments, taking measurements, or performing technical tasks, attending to special cases or exceptions defined in the text. (RST.9-10.3)

Craft and Structure

- Determine the meaning of symbols, key terms, and other domain-specific words and phrases as they are used in a specific scientific or technical context relevant to grades 9-10 texts and topics. (RST.9-10.4)

Integration of Knowledge and Ideas

- Translate quantitative or technical information expressed in words in a text into visual form (e.g., a table or chart) and translate information expressed visually or mathematically (e.g., in an equation) into words. (RST.9-10.7)

Range of Reading and Level of Text Complexity

- By the end of grade 10, read and comprehend science/technical texts in the grades 9-10 text complexity band independently and proficiently. (RST.9-10.10)

Writing

Research to Build and Present Knowledge

- Conduct short as well as more sustained research projects to answer a question (including a self-generated question) or solve a problem; narrow or broaden the inquiry when appropriate; synthesize multiple sources on the subject, demonstrating understanding of the subject under investigation. (W.9-10.7)
- Draw evidence from literary or informational texts to support analysis, reflection, and research. (W.9-10.9)

Speaking and Listening

Comprehension and Collaboration

- Initiate and participate effectively in a range of collaborative discussions (one-on-one, in groups, and teacher-led) with diverse partners on grades 9-10 topics, texts, and issues, building on others' ideas and expressing their own clearly and persuasively. (SL.9-10.1)
- Integrate multiple sources of information presented in diverse media or formats (e.g., visually, quantitatively, orally) evaluating the credibility and accuracy of each source. (SL.9-10.2)

Modeling Bird Population Trends

Before You Start...

Time

Preparation: **15 minutes**

Instruction: **90 minutes**

Preparation

- Make copies of the Student Data Table, Reading, and Worksheets.
- Download the *"Birds 4 Teacher Slides"* PowerPoint file.

Technology

A-V Equipment

- LCD Projector
- Screen or Whiteboard

Computers

- Internet Connection

Information and Communication Technology

- eBird animated maps online - http://crossingboundaries.org/bwb.php

Low-Tech Options

- Use PowerPoint file and/or printed copies of the maps at various points throughout a year.

Materials

Student Reading

- Mapping and Modeling with eBird Citizen Science Data

Worksheets

- Student Data Table
- Maps Produced Through Modeling: Animated "Occurrence Maps"
- Mapping and Modeling with eBird Citizen Science Data

Modeling Bird Population Trends
The Investigation

Investigation Overview

1. Introduce an animated map showing changes in locations of Indigo Buntings in the U.S. across the seasons.

2. Relate bird migration to landscape features.

3. Compare and contrast animated maps showing migration patterns of four related bird species.

4. Read and reflect on mapping and modeling using citizen science data.

Learning Objectives

Students will be able to:

- Interpret map sequences to draw conclusions about seasonal distribution and habitat needs of individual bird species.

Key Concepts

- Life cycles
- Population dynamics
- Migration
- Citizen science
- Modeling

Technology Overview

- Students use web-based animated maps.

Conducting the Investigation

1. Introduce an animated map showing changes in locations of Indigo Buntings in the U.S. across the seasons.

   a. Without explaining anything about what it portrays, show the eBird animated "Occurrence Map" for Indigo Buntings (Slide #2). Watch it cycle through the seasons for one or more years (it will repeatedly cycle through a year). Ask students to think about what this animated sequence of maps could be portraying.

   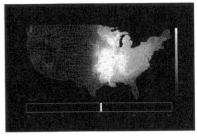
   Slide 2

   > *After some discussion, explain that the map lights up to show presence of a bird species called the Indigo Bunting. The brighter the lit-up area, the higher the probability of seeing one or more Indigo Buntings if you were to go birding at each particular location and time of year.*

   b. Have students work in small groups to brainstorm what is happening as part of the map lights up during the middle of the year and then darkens again. Reconvene to discuss ideas.

   c. Have each student draft an individual response to Question 1 on the **Maps Produced Through Modeling: Animated "Occurrence Maps" Worksheet**.

   > *The progression in lighting shows movement of this species into the U.S. from the south, gradually spreading northward throughout*

the eastern half of the country. The scale across the bottom indicates that this movement begins in the spring, intensifies in early summer, and then reverses in the fall. Prompt students to think about possible biological interpretations of these patterns, eventually leading to discussion of migration of this species northward to the U.S. for breeding and then back south for the winter months.

2. Relate bird migration to landscape features.

    a. Show a map of landscape features in the U.S. (Slide #3). Ask students whether any of the major features such as the Mississippi River Basin or the Rocky or Appalachian Mountain Ranges correspond with areas they see lighting up on the animated map. Give them time to write a response to Question 2 on the worksheet.

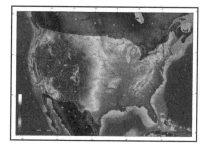
Slide 3

    b. Discuss their responses to Question 2, and collectively think about the different types of habitats provided by the various types of landscapes seen on the U.S. map. Point out that Indigo Buntings migrate into the U.S. to breed in the summer, and remind students that they explored the importance of specific types of habitat for bird nesting in Investigation 2 focusing on nesting birds in New York State.

    Indigo Buntings arrive on the Gulf Coast in late the spring after migrating north from Central America and the Caribbean. They spread northward through the Mississippi River Valley and disperse across most of the eastern U.S. for nesting season. Looking closely at distribution during the summer months, you can see dark spots surrounded by brighter zones. These dark spots are urban areas, dark because Indigo Buntings do not breed in cities. In the fall, they head south to winter in warmer locations.

3. Compare and contrast animated maps showing migration patterns of four related bird species.

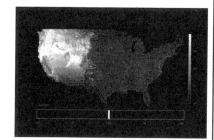
Slide 4

    a. Have students work individually or in small groups to view the animated maps for three additional species and complete Questions 3 and 4 on the worksheet:

    - Lazuli Bunting (Slide #4) http://tinyurl.com/pral93r
    - Painted Bunting (Slide #5) http://tinyurl.com/o58o5xq
    - Northern Cardinal (Slide #6) http://tinyurl.com/q3dcmb3

    b. Discuss student findings in terms of migratory patterns for all four species.

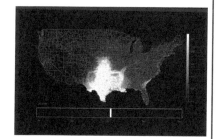
Slide 5

    Although all three of these bunting species are in the same family, they show markedly different migration patterns. See Teacher Notes for more information about these species.

    Unlike the other three species, the Northern Cardinal does not migrate. However, students may have noticed that the map does show slight variation from season to season. Although these birds can be seen year-round, they are harder to detect when they are

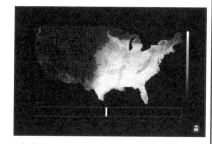
Slide 6

sitting quietly on their nests to protect their young than at other times of year. In winter, cardinals can readily be seen at bird feeders as well as in the wild. In the spring, they flaunt their bright plumage and vibrant songs to attract mates. Cardinals are harder to see once nesting begins because then they are sitting quietly on their nests to protect their young.

4. Read and reflect on mapping and modeling using citizen science data.

   a. Have students read ***Mapping and Modeling with eBird Citizen Science Data*** and respond to the questions on the **Mapping and Modeling with eBird Citizen Science Data Worksheet**.

   b. As a class, discuss the fact that the models that produced these animated maps rely on bird observation data submitted to the eBird citizen science project by many thousands of volunteers, including students. Discuss the relationship between data and modeling and ask, "Considering the huge wealth of bird observation data entered into eBird by citizen scientists, why is it useful to model bird population dynamics rather than just map the eBird data points?"

   > *The models integrate the bird observation data with data on land use, climate, and other environmental variables related to habitat. Whereas the eBird data show exactly where a species has been sighted, the models make it possible to predict what has happened in the past or will happen in the future, and to fill in gaps across the landscape where insufficient bird data exist.*

   c. Tell the students that in the next Investigation you will be using eBird data to further explore the distributions of the four species you have looked at here.

Selected Species Notes from *eBird* and *All About Birds*

Northern Cardinal

Distribution and Migration. Northern Cardinal is one of the most familiar birds in the Lower 48 states, seen from Maine to Florida and west to Arizona and Colorado. It is not known to have any significant migrations, although they do perform short distance movements and may gather in flocks in winter.

Habitat. Dense shrubby areas such as forest edges, overgrown fields, hedgerows, backyards, marshy thickets, mesquite, regrowing forest, and ornamental landscaping. Cardinals nest in dense foliage. Growth of towns and suburbs has helped this species expand its range northward.

Food. Seeds and fruit, supplemented with insects (and feeding nestlings mostly insects). Common fruits and seeds include dogwood, wild grape, buckwheat, grasses, sedges, mulberry, hackberry, blackberry, sumac, tulip-tree, and corn. Cardinals eat many kinds of birdseed, particularly black oil sunflower seed.

http://ebird.org/content/ebird/?p=980 and http://www.allaboutbirds.org/guide/northern_cardinal/lifehistory

Indigo Bunting

Distribution and Migration. In spring, Indigo Buntings arrive on the Gulf Coast after crossing from the Yucatan. They surge north through the Mississippi River Valley. They head south back across and around the Gulf of Mexico primarily during October.

Habitat. Weedy and brushy areas. They love edges where fields meet forests. When not singing from tall perches, they can often be seen foraging among seed-laden shrubs and grasses. While migrating and in winter, they forage in fields, lawns, grasslands, rice fields, as well as in shrubs, and trees.

Food. Indigo Buntings eat small seeds, berries, buds, and insects. Spiders and insect prey form the majority of their diet during summer months. The brown-tail moth caterpillar, which is covered with noxious hairs that cause nasty rashes and respiratory problems in people, presents no obstacle to a hungry bunting.

http://ebird.org/content/ebird/?p=954 and http://www.allaboutbirds.org/guide/Indigo_Bunting/lifehistory

Lazuli Bunting

Distribution and Migration. In the spring, Lazuli Buntings stream north through Arizona and into California and the Rocky Mountains. In the fall, they congregate in the Mojave Desert before heading to Central America for the winter.

Habitat. Lazuli Buntings are common in shrubby areas throughout the American West. Their habitats include brushy hillsides, riparian habitats, wooded valleys, sagebrush, chaparral, open scrub, recent post-fire habitats, thickets and hedges along agricultural fields, and residential gardens.

Food. Lazuli Buntings forage for seeds, fruits, and insects, and they also eat at bird feeders.

http://ebird.org/content/ebird/?p=967 and http://www.allaboutbirds.org/guide/Lazuli_Bunting/lifehistory

Painted Bunting

Distribution and Migration. There are two distinct populations. The more widespread subspecies is found in Texas, eastern New Mexico, Kansas, Oklahoma, Arkansas, and Louisiana, and these birds winter in west Mexico and south Florida. The eastern subspecies breeds in northeastern Florida, Georgia, South Carolina, and southern North Carolina and spend the winters in south Florida, the Bahamas, and Cuba.

Habitat. Painted Buntings breed in semi-open habitats with scattered shrubs or trees. In winter they prefer high grass, shrubby overgrown pasture, and thickets.

Food. Painted Buntings eat seeds for most of the year, switching to mostly insects in the breeding season. They forage on the ground for seeds and also often come to feeders. During the breeding season they may pull invertebrates from spider webs, or even dive straight through a web to steal a spider's prey.

Conservation Status. Painted Buntings are classified by the IUCN as "Near Threatened" and Partners in Flight has designated them a species of concern. They are still fairly common, but their populations are dropping as urban development causes habitat loss and degradation. Another threat is trapping of Painted Buntings on their wintering grounds and for illegal sale as pets, particularly in Mexico and the Caribbean.

http://ebird.org/content/ebird/?p=962 and http://www.allaboutbirds.org/guide/Painted_Bunting/lifehistory

Mapping and Modeling with eBird Citizen Science Data
Student Reading

What is eBird, and how can it help us better understand bird population dynamics?

eBird is a huge and rapidly growing citizen science project. *Citizen science* refers to efforts in which volunteers partner with professional scientists to collect or analyze data. In eBird, any person anywhere in the world can submit data about birds they have seen. Collectively, these efforts are building a database that offers insights into the distribution, abundance, and population trends for most of the world's over 10,000 bird species.

How can you use eBird?

If you set up an account in eBird and enter data about birds you have seen, you can use eBird to store and keep track of your sightings. Whether or not you enter any data of your own, you can use eBird's "Explore Data" functions to determine what birds live in your area (or any area of interest), locate the best places in which to go birding, and discover what rare or unusual species have been seen.

With bar charts and line graphs generated by eBird, you can explore what birds live in a state, county, or other area of interest, which species migrate, when they arrive and depart, and how many are seen at various times of year.

Using eBird's interactive Range and Point maps, you can select a species and view where it has been reported throughout the world. Zooming in, you can see the details at any particular location.

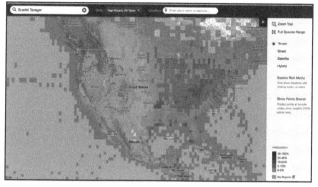

eBird Range and Point Map for Scarlet Tanager

How is eBird used by professional scientists

Scientists use eBird data to follow trends and conduct research on bird populations. Using complex mathematical models that run on supercomputers, scientists also are integrating eBird data with dozens of other variables that collectively describe habitat. These models generate the animated "Occurrence Maps" that show predicted patterns of movement for individual bird species. These maps show the *probability of occurrence* for a species each week at more than 10,000 locations across the lower 48 states. They are called "occurrence maps" because they show the likelihood of seeing that species in each location and time of year. The models currently show results only within the continental U.S. because of limitations in compatible data describing land use beyond these borders.

With so much data showing where birds have been seen, why bother with models?

One reason for creating models is to fill the gaps where data are lacking. eBird has an impressive and ever-growing database, but the data come primarily from places where people like to go birding. Modeling helps to fill the gaps between these locations, predicting where species are likely to occur even when nobody has reported from those places yet. Modeling also makes it possible to explore change over time. The animated maps show the migration patterns of each species and are helping scientists to discover species-specific and habitat needs throughout the year. Another reason for modeling is to make predictions into the future. For example, adjusting the climate variables in the model make it possible for scientists to investigate how the timing of migration might be affected for each species.

On a broad scale, the eBird occurrence maps show migration patterns for individual bird species. Zooming in, scientists can see details of the types of habitat used by each species during various times of year. These details are being used to understand ecological relationships and to plan conservation measures. For example, eBird modeling is used to identify priorities for land conservation to protect habitats needed by specific types of birds. For more information, see the *State of the Birds* reports (http://stateofthebirds.org/).

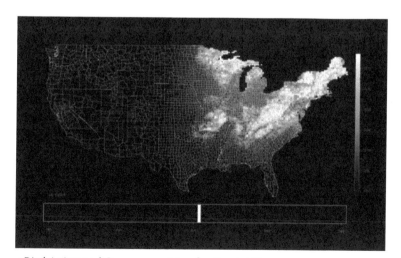

eBird Animated Occurrence Map for Scarlet Tanager

Name: _____ Period: _____ Date: _____

Data Table

	Instructions	Indigo Bunting	Lazuli Bunting	Painted Bunting	Northern Cardinal
A	Draw a picture, insert a photo, or use words to describe what each of these bird species looks like.				
B	Using eBird animated "Occurrence Maps," shade in the area or insert a re-sized screen shot to show the peak distribution of each species in the U.S.				
C	Using eBird "Range and Point" maps, shade in the map or insert a re-sized screen shot showing the year-round distribution of each species.				
D	Using Range Maps from All About Birds (www.allaboutbirds.org) and observe the range maps color-coded by season and shade in (or insert an appropriate graphic to) the map.				

Data Table Answer Key

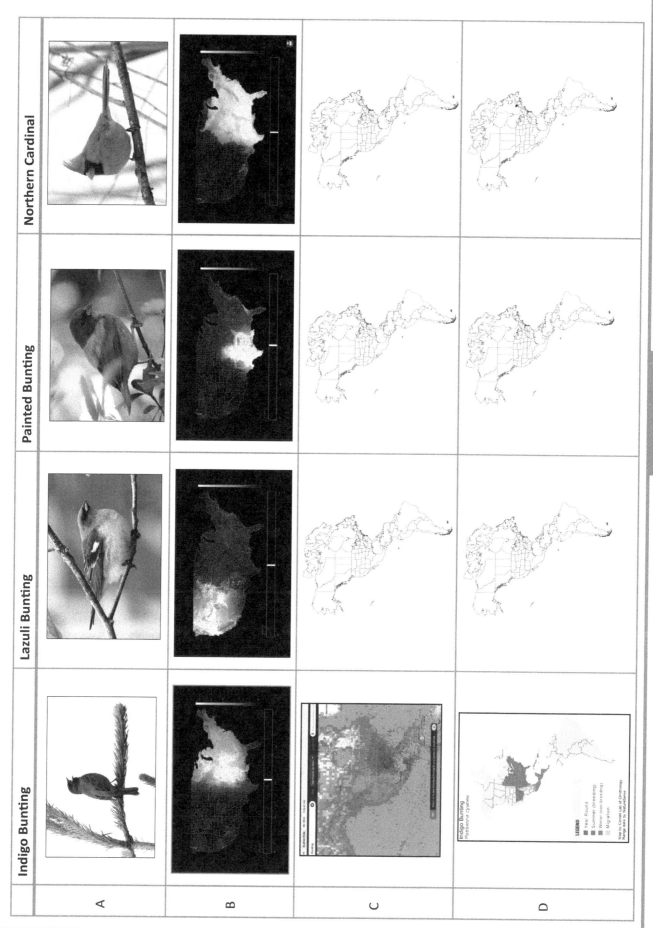

	Indigo Bunting	Lazuli Bunting	Painted Bunting	Northern Cardinal
A				
B				
C				
D				

Maps Produced through Modeling: Animated "Occurrence Maps" Worksheet

Investigation 4

1. Watch the Indigo Bunting map cycle through the seasons (http://tinyurl.com/lq8ojyn) and think about why the map lights up and then darkens as you respond to these questions:

    a. At what times of year does the map light up?

    b. Why does the map light up in some parts of the country and not others? (What is happening in the bright regions?)

    c. What biological phenomenon might explain these seasonal changes?

2. What major landscape features correspond with the areas that light up on the map? What patterns do you notice?

3. Take a look at the animated maps for three more bird species:

- Lazuli Bunting: http://tinyurl.com/pral93r
- Painted Bunting: http://tinyurl.com/o58o5xq
- Northern Cardinal: http://tinyurl.com/q3dcmb3

   a. In Row A of the Data Table, draw a simple picture or insert a photo or write a brief description of the characteristics of each of these species.

   b. In Row B, sketch the extent of the area that lights up for each species, or insert a resized screenshot taken at the time when the maximum area of the country is lit up.

4. Looking at the four maps in Row B, what can you conclude about the parts of the country in which these four species build their nests and raise their young?

Indigo Bunting:

Lazuli Bunting:

Painted Bunting:

Northern Cardinal:

Maps Produced through Modeling: Animated "Occurrence Maps"
Worksheet Answer Key

1. Watch the Indigo Bunting map cycle through the seasons (http://tinyurl.com/lq8ojyn) and think about why the map lights up and then darkens as you respond to these questions:

   a. At what times of year does the map light up?

   In the spring and summer. It begins lighting up in April, with the lit-up areas getting brighter and more widespread through the summer months and then tapering off in the fall.

   b. Why does the map light up in some parts of the country and not others? (What is happening in the bright regions?)

   The bright areas indicate where Indigo Buntings are likely to be located at each time of year. The map does not light up in regions where this species is not found.

   c. What biological phenomenon might explain these seasonal changes?

   Indigo Buntings are migrating into and across the Eastern U.S. to nest and breed in summer months. They fly south in the fall to their wintering grounds beyond the borders of this map.

2. What major landscape features correspond with the areas that light up on the map? What patterns do you notice?

   In spring, Indigo Buntings arrive on shores along the Gulf of Mexico and spread north through the Mississippi River Valley, eventually covering most of the eastern U.S. In the fall, they gather again in the Mississippi River Valley and along the coast before heading farther south for the winter.

3. Take a look at the animated maps for three more bird species:

- Lazuli Bunting: http://tinyurl.com/pral93r
- Painted Bunting: http://tinyurl.com/o58o5xq
- Northern Cardinal: http://tinyurl.com/q3dcmb3

a. In Row A of the Data Table, draw a simple picture or insert a photo or write a brief description of the characteristics of each of these species.

b. In Row B, sketch the extent of the area that lights up for each species, or insert a resized screenshot taken at the time when the maximum area of the country is lit up.

See Teacher Notes (page 7) for sample photos and maps.

4. Looking at the four maps in Row B, what can you conclude about the parts of the country in which these four species build their nests and raise their young?

Indigo Bunting:

Primarily in the eastern U.S.

Lazuli Bunting:

Primarily in the eastern U.S.

Painted Bunting:

Primarily in the eastern U.S.

Northern Cardinal:

Primarily in the eastern U.S.

Mapping and Modeling with eBird Citizen Science Data
Worksheet

1. What is citizen science?

2. What kind of data is collected in eBird?

3. Who submits eBird data, and from what parts of the world?

4. How can you use eBird data?

5. How do professional scientists use eBird data?

6. eBird is collecting huge amounts of data from all over the world. With so much bird observation data available, why is it still useful to model where each species is likely to occur?

Mapping and Modeling with eBird Citizen Science Data
Worksheet Answer Key

1. What is citizen science?

    Efforts in which volunteers partner with professional scientist to collect or analyze data.

2. What kind of data is collected in eBird?

    Bird observation data. Volunteers submit data about the bird species they have seen.

3. Who submits eBird data, and from what parts of the world?

    Anyone can submit data to eBird about the bird species they have seen anywhere in the world.

4. How can you use eBird data?

    You can use it to keep track of your own bird sightings, and you can explore data entered by anyone to figure out what birds have been seen in any area of interest. You can use the data exploration tools to figure out which species migrate, when they arrive and depart, and how many are seen at various times of year.

5. How do professional scientists use eBird data?

    They use eBird data to follow trends and conduct research on bird populations, including creating models that predict movement of each species over the course of a year.

6. eBird is collecting huge amounts of data from all over the world. With so much bird observation data available, why is it still useful to model where each species is likely to occur?

    Data collected through eBird show the exact times and locations in which people have reported seeing each species. Modeling makes it possible to go beyond these observation points and predict when and where each species is expected to occur. The models can be used to make predictions into the future and into locations with no eBird participants.

5
investigation

Tracking Birds
with Citizen Science

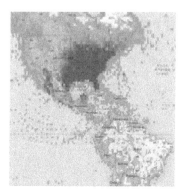

Investigation 5 Photo Credits

Investigation 5
Tracking Birds with Citizen Science

In this Investigation, students use eBird citizen science data to explore the distribution of four related bird species over time and space, and they consider the roles of citizen science and modeling in scientific research.

Essential Question

- How are citizen science data and modeling useful in tracking bird population dynamics?

Science Concepts and Topics

In eBird, citizen scientists submit checklists detailing what bird species they have seen at a specified time and location. Using the assembled data, scientists have created models that produce the animated maps used in Investigation 4. In this current Investigation, students work more directly with the eBird citizen science data. Using web-based data visualization tools, they view maps and create graphs representing times and places in which individual species have been sighted by participating citizen scientists.

Standards Addressed in This Investigation

Next Generation Science Standards – Middle School

- Construct an explanation that predicts patterns of interactions among organisms across multiple ecosystems. (MS-LS2-2)

Next Generation Science Standards – High School

- Use mathematical representations to support and revise explanations based on evidence about factors affecting biodiversity and populations in ecosystems of different scales. (HS-LS2-2)

Next Generation Science Standards – Scientific Practices

- Asking questions and defining problems
- Planning and carrying out investigations
- Analyzing and interpreting data
- Using mathematics and computational thinking
- Constructing explanations and designing solutions
- Engaging in argument from evidence
- Obtaining, evaluating, and communicating information

Common Core (9-10ᵗʰ grade)

Reading Science & Technical Subjects

Key Ideas and Details

- Follow precisely a complex multistep procedure when carrying out experiments, taking measurements, or performing technical tasks, attending to special cases or exceptions defined in the text. (RST.9-10.3)

Craft and Structure

- Determine the meaning of symbols, key terms, and other domain-specific words and phrases as they are used in a specific scientific or technical context relevant to grades 9-10 texts and topics. (RST.9-10.4)

Integration of Knowledge and Ideas

- Translate quantitative or technical information expressed in words in a text into visual form (e.g., a table or chart) and translate information expressed visually or mathematically (e.g., in an equation) into words. (RST.9-10.7)

- Compare and contrast findings presented in a text to those from other sources (including their own experiments), noting when the findings support or contradict previous explanations or accounts. (RST.9-10.9)

Writing

Research to Build and Present Knowledge

- Conduct short as well as more sustained research projects to answer a question (including a self-generated question) or solve a problem; narrow or broaden the inquiry when appropriate; synthesize multiple sources on the subject, demonstrating understanding of the subject under investigation. (W.9-10.7)

- Gather relevant information from multiple authoritative print and digital sources, using advanced searches effectively; assess the usefulness of each source in answering the research question; integrate information into the text selectively to maintain the flow of ideas, avoiding plagiarism and following a standard format for citation. (W.9-10.8)

- Draw evidence from literary or informational texts to support analysis, reflection, and research. (W.9-10.9)

Speaking and Listening

Comprehension and Collaboration

- Initiate and participate effectively in a range of collaborative discussions (one-on-one, in groups, and teacher-led) with diverse partners on grades 9-10 topics, texts, and issues, building on others' ideas and expressing their own clearly and persuasively. (SL.9-10.1)

- Integrate multiple sources of information presented in diverse media or formats (e.g., visually, quantitatively, orally) evaluating the credibility and accuracy of each source. (SL.9-10.2)

Tracking Birds with Citizen Science
Before You Start...

Time

Preparation: **15 minutes**

Instruction: **90 minutes**

Preparation

- Make copies of the Student Data Table and Worksheet.

- Download the *"Birds 5 Teacher Slides"* PowerPoint file.

Technology

A-V Equipment

- LCD Projector
- Screen or Whiteboard

Computers

- Internet Connection

Information and Communication Technology

- eBird web-based data visualization tools maps

Low-Tech Options

- Use PowerPoint file and/or printed copies of the maps and graphs.

Materials

Worksheets

- Student Data Table (started in Investigation 4)
- Mapping eBird Data Worksheet

Investigation 5

Tracking Birds with Citizen Science
The Investigation

Investigation Overview

1. Use eBird maps to view the distribution of sightings of the four bird species introduced in Investigation 4.

2. Explore other dimensions of bird population dynamics using eBird's "Line Graph" functions.

3. Conclude by discussing the roles of citizen science and modeling in understanding bird population distribution and dynamics.

Learning Objectives

Students will be able to:

- Interpret maps to compare the distributions of selected bird species.

- Describe the differences in meaning between maps produced through modeling versus those portraying citizen science data.

Key Concepts

- Population dynamics
- Migration
- Citizen science
- Modeling

Technology Overview

- Students use use web-based data visualization tools provided by the eBird citizen science project.

Conducting the Investigation

1. Use eBird maps to view the distribution of sightings of the four bird species introduced in Investigation 4.

    a. Project Slide #2 to remind students about the animated maps used in Investigation 4, and discuss how these maps relate to citizen science.

    Slide 2

    These maps show the results of models to portray changes in bird populations throughout a year. The models use data describing a wide range of environmental variables, and they also incorporate data that have been submitted to the eBird citizen science project.

    b. Explain that now we will switch from looking at the model outputs to looking at visualizations of the eBird data itself. Project the eBird website (http://ebird.org/ or Slide #3) and ask students to recall from their Investigation 4 reading what eBird is and how it works.

    Slide 3

    In eBird, anyone anywhere in the world can submit data about birds they have observed. Collectively, these observations by citizen scientists are creating a massive database of use to professional scientists, birders, students, and anyone interested in exploring distribution of almost any of the world's over 10,000 species of birds.

    Slide 4

c. Under the eBird's *Explore Data* tab, select *Range and Point Maps*. On the map page that appears, type "Indigo Bunting" into the search box. (Use Slides 4 and 5 if Internet access is unavailable.)

Slide 5

d. The resulting map will look like the one in Row C of the students' **Data Sheet**. Discuss what this map shows and how it differs from the animated map for this same species (refer back to Slide #2 if needed).

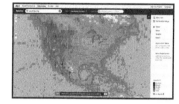
Slide 6

> *Note that some areas are shaded purple on the eBird map. These indicate places where eBird participants have reported seeing one or more Indigo Buntings. The darker the purple, the greater the percentage of checklists that included Indigo Buntings. If you zoom in far enough, you can click on markers that show the full list of bird species reported by each individual, along with the date and timing of each checklist.*

> *Some areas are shaded grey. Each grey square indicates a place in which eBird participants have submitted checklists that did not include Indigo Buntings.*

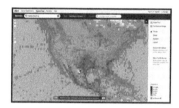
Slide 7

e. Together as a class, or at student computer workstations, have students take a look at the eBird "Range and Point" maps for the other three species. (Go to http://ebird.org/ebird/map/ and type in each species name in the search bar on this page, or use Slides 6-8 for non-real-time versions of these maps).

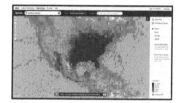
Slide 8

f. Have students complete Questions 1-3 on the **Mapping eBird Data Worksheet**, then reconvene to discuss their findings.

> *If you compare the maps for these four species in the Cardinalidae family, you can see that the purple shading roughly corresponds with the areas that were brightly lit up in mid-summer in the animated maps viewed in the previous Investigation. (Compare rows B and C in the Teacher Notes or the students' Data Table).*

> *It might be tempting to think that the purple shading illustrates the migratory range of each species. However, the eBird data map represents year-round sightings, not movement. Using the "Date" feature, you can select specific months rather than the year-round default setting, making it possible to compare distribution of the selected species from one season to another and determine whether or not it migrates between nesting and wintering grounds.*

Slide 9

> *A simpler way to look at migratory status of each species is to use range maps such as those shown in Row D (Slide #9). Colored shading in these maps indicate broad regions in which each species lives year-round or during summer, winter, and migratory seasons. These maps are generalizations rather than showing specific data points.*

2. Explore other dimensions of bird population dynamics using eBird's **Line Graph** functions.

   a. Explain that for each species, the eBird website will create graphs using the variables listed below. You can designate the species, geographic area, and timeframe of interest. It is possible to graph several species on a single graph or to combine several variables for a single species.

      • Frequency is the % of checklists reporting that species within a specified date range and region.

      • Abundance is the average # birds in that species that have been reported on all checklists within a specified date range and region. These data tell us what we might expect to see when going out birding on an average day.

      • Birds per hour is the average # birds in that species seen per hour spent birding within a specified date range and region.

      • Average Count is the average number of birds in that species seen on checklists with a positive observation for the species within a specified date range and region. (This differs from Abundance in that it only incorporates checklists that reported the species, essentially telling us how many of each species we can expect to see where the species is encountered.)

      • High Count is the highest count of the species submitted on a single checklist within a specified date range and region.

      • Totals is the sum of all observations of the species from all checklists submitted within a specified date range and region.

   b. Together as a class, create an example line graph for a species of interest. For example, the graph in Slide # 10 shows frequency of sightings of Bald Eagles in North America from 1900-2014. If you produce this graph on the eBird website, you can use the "Change Location" button to select your state or narrow down to your county. Comparing this new graph with the earlier one, you can compare average annual distribution of Bald Eagles in your area to that in North America as a whole.

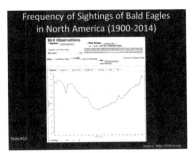

Slide 10

      Note that you begin the graphing function by selecting one or more bird species. The first graph you will see represents sighting data starting in 1900 for your selected species across all of North America. Using buttons that appear above this graph, you can narrow down the dates and location of interest. For dates, you can select either a range of years to view cumulatively or up to five consecutive years to view individually. For location, you can select your state and county or other types of region.

c. Collectively brainstorm the kinds of questions that would be interesting to address by graphing one or more of the eBird variables.

> *For example, one option is to select a species in your area that migrates and another that does not, then create a line graph combining these two species to compare frequency of sightings over the course of a year. Abundance or high counts could also be interesting to look at, to see whether your selected species congregate in groups during any particular season.*

> *Another possibility is to investigate how consistent one of these variables is from year to year for a selected bird species. In this case, you would create a graph showing multiple years worth of data for a single species. The 2015 State of the Birds Report listed 33 common bird species that are rapidly declining in many areas. Students could download that list and investigate abundance of one or more of those species in your community (http://www.stateofthebirds.org/ abundance).*

d. Have students explore eBird's "Line Graph" functions and identify a question to address by graphing and interpreting one or more of the possible variables. The most likely approach is for students to generate questions of comparison. They might compare the same species at two different locations. Or compare two different species at the same location, and so on.

e. Ask students to print their graphs or save screenshots for display.

f. Display the students' graphs, and discuss their findings in terms of any trends they found in bird population dynamics over the course of a year or multiple years.

3. Conclude by discussing the roles of citizen science and modeling in understanding bird population distribution and dynamics.

> *The eBird citizen science project is collecting a wealth of data on distribution of bird species throughout the world. Collectively, the armies of birders who contribute their data are assembling data far beyond what professional scientists could accomplish on their own.*

> *These data are shown on the web in the form of maps, charts, and graphs that indicate where and when each species has been seen.*

> *Combining eBird data with a wide range of environmental variables, scientists are creating models that predict the likelihood of each bird species appearing at a particular place and time. These produce the animated maps used in Investigation 4.*

> *Modeling makes it possible to extend beyond the citizen science data to reach conclusions about places and times for which no bird*

observation data exist. For example, predictions can be made about the effects of future climate change on the summer or winter range of a selected bird species, or on the dates when this species would be expected to arrive at or depart from its nesting grounds.

4. Possible extension:

    a. Invite students to look for birds and enter their sightings in eBird. Your class can register collectively for an eBird account, or each student can have their own account. If you have the opportunity to take your class outside, or even to watch birds out the window, you could conduct a short bird count together and demonstrate how to enter the data online. For more information, download the **Using eBird With Groups** guide: http://dl.allaboutbirds.org/usingebirdwithgroups

Data Table

	Instructions	Indigo Bunting	Lazuli Bunting	Painted Bunting	Northern Cardinal
A	Draw a picture, insert a photo, or use words to describe what each of these bird species looks like.				
B	Using eBird animated "Occurrence Maps," shade in the area or insert a re-sized screen shot to show the peak distribution of each species in the U.S.				
C	Using eBird "Range and Point" maps, shade in the map or insert a re-sized screen shot showing the year-round distribution of each species.				
D	Using Range Maps from All About Birds (www.allaboutbirds .org) and observe the range maps color-coded by season and shade in (or insert an appropriate graphic to) the map.				

Data Table Answer Key

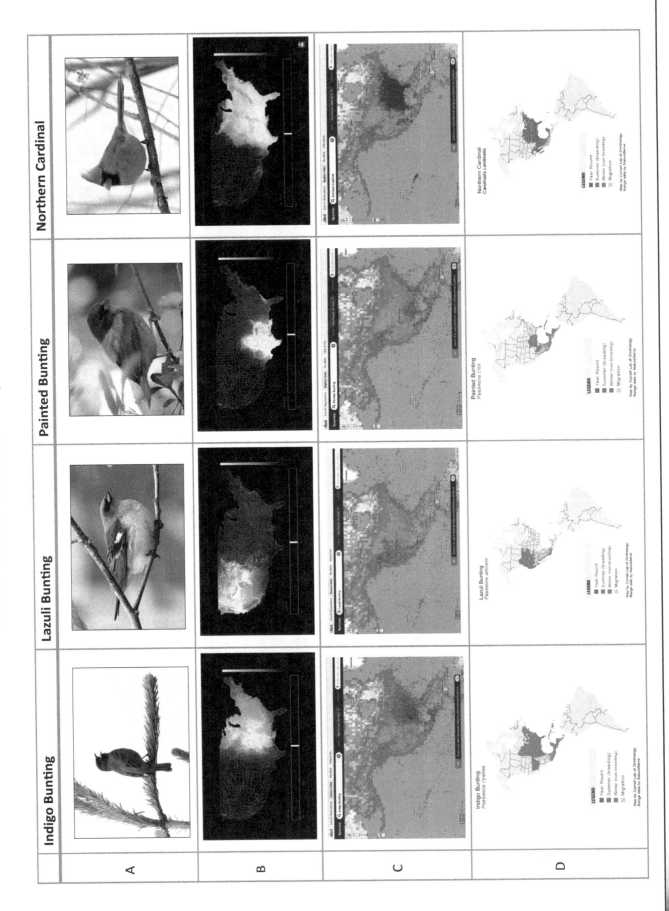

Mapping eBird Data
Worksheet

On the eBird website (http://ebird.org/), select the **Explore Data** tab, then select **Range and Point Maps**. On the map page that appears, type "Indigo Bunting" into the search box. You should see a map like the one in Row C of your Data Table. In the bottom right corner of this map is the legend:

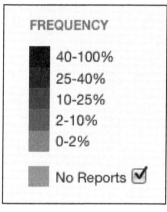

This tells us that the darker the purple on the map, the greater the frequency of sighting of Indigo Buntings. This species is present but more rarely seen in the areas with lighter purple boxes. The grey "No Reports" boxes indicate areas in which eBird participants have submitted bird checklists that did not include Indigo Buntings. In areas with no purple or grey shading, no data have been submitted yet to eBIrd.

1. In Row C of your Data Table, use the Range and Point Map for each species to color the maps (or insert re-sized screenshots) to show where each species has been reported by eBird participants.

2. Compare the eBird data maps in Row C with the modeling maps in Row B.

   a. How do the lit-up regions in Row B maps compare with the purple areas in Row C?

   b. How do these two types of maps differ in terms of the types of information they portray?

3. Sketch or insert range maps in Row D of the Data Table, using the All About Birds website (www.allaboutbirds.org) to look up each of the selected species.

a. Looking across Row D, what sets the Northern Cardinal apart from the other three species?

b. How do your maps for Row C and Row D compare? What differences do you see, and what might be causing these differences?

4. Which row of maps does each of these descriptions represent?

-   These maps show generalized regions where the species is likely to occur in various seasons:

-   These maps show where the species has actually been seen and reported by citizen scientists:

-   These maps show the results of modeling that indicates where the species is expected to occur:

Mapping eBird Data
Worksheet Answer Key

On the eBird website (http://ebird.org/), select the **Explore Data** tab, then select **Range and Point Maps**. On the map page that appears, type "Indigo Bunting" into the search box. You should see a map like the one in Row C of your Data Table. In the bottom right corner of this map is the legend:

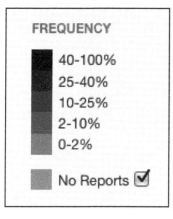

This tells us that the darker the purple on the map, the greater the frequency of sighting of Indigo Buntings. This species is present but more rarely seen in the areas with lighter purple boxes. The grey "No Reports" boxes indicate areas in which eBird participants have submitted bird checklists that did not include Indigo Buntings. In areas with no purple or grey shading, no data have been submitted yet to eBIrd.

1. In Row C of your Data Table, use the Range and Point Map for each species to color the maps (or insert re-sized screenshots) to show where each species has been reported by eBird participants.

2. Compare the eBird data maps in Row C with the modeling maps in Row B.

   a. How do the lit-up regions in Row B maps compare with the purple areas in Row C?

   They cover similar regions.

   b. How do these two types of maps differ in terms of the types of information they portray?

   The Range and Point maps in Row C show the exact times and locations in which eBird participants have reported seeing each species. The Occurrence Maps in Row B show the expected probability of occurrence of each species based on modeling of eBird data along with data for variables that describe habitat conditions.

3. Sketch or insert range maps in Row D of the Data Table, using the All About Birds website (www.allaboutbirds.org) to look up each of the selected species.

Investigation 5

a. Looking across Row D, what sets the Northern Cardinal apart from the other three species?

The Northern Cardinal is the only one of these species that stays in the U.S. year round. The other three species migrate south for the winter.

b. How do your maps for Row C and Row D compare? What differences do you see, and what might be causing these differences?

The maps are similar for each species, but the eBird Range and Point maps in Row C show sightings in some areas that are not colored in the Row D maps. The Row D maps show generalized regions in which each species is expected to occur, whereas the Row C maps show actual locations in which these species have been sighted. This includes rare sightings, for example when a bird has been swept off course by a hurricane.

4. Which row of maps does each of these descriptions represent?

- These maps show generalized regions where the species is likely to occur in various seasons:

  Row D

- These maps show where the species has actually been seen and reported by citizen scientists:

  Row C

- These maps show the results of modeling that indicates where the species is expected to occur:

  Row B

6 investigation

Investigating Bird Biodiversity Across the Americas

Investigation 6 Photo Credits

Investigation 6
Investigating Bird Biodiversity Across the Americas

In this Investigation, students explore bird species richness in the U.S. and across the Western Hemisphere. They create graphs and use maps, quantitative data, and web resources to investigate distribution of selected bird species in the U.S. and beyond.

Essential Question

- How does bird species richness compare between the U.S. and other countries in North and South America?

Science Concepts and Topics

Species richness, the number of species in a country or other designated area, is highest along the Equator and declines toward the poles. This *latitudinal gradient* applies to plant as well as animal species. Consistently moderate temperatures and abundant rainfall in the tropics support abundant vegetation, and the high primary productivity supports a wealth of animal life. Diverse habitats in the tropics support *high biodiversity*, including roughly one third of the world's species of birds.

Standards Addressed in This Investigation

Next Generation Science Standards – Middle School

- Analyze and interpret data to provide evidence for the effects of resource availability on organisms and populations of organisms in an ecosystem. (MS-LS2-1)
- Construct an explanation that predicts patterns of interactions among organisms across multiple ecosystems. (MS-LS2-2)

Next Generation Science Standards – High School

- Use mathematical representations to support and revise explanations based on evidence about factors affecting biodiversity and populations in ecosystems of different scales. (HS-LS2-2)

Next Generation Science Standards – Scientific Practices

- Asking questions and defining problems
- Planning and carrying out investigations
- Analyzing and interpreting data
- Using mathematics and computational thinking
- Constructing explanations and designing solutions
- Engaging in argument from evidence
- Obtaining, evaluating, and communicating information

Investigation 6

Common Core (9-10th grade)

Reading Science & Technical Subjects

Key Ideas and Details

- Cite specific textual evidence to support analysis of science and technical texts, attending to the precise details of explanations or descriptions. (RST.9-10.1)

- Determine the central ideas or conclusions of a text; trace the text's explanation or depiction of a complex process, phenomenon, or concept; provide an accurate summary of the text. (RST.9-10.2)

- Follow precisely a complex multistep procedure when carrying out experiments, taking measurements, or performing technical tasks, attending to special cases or exceptions defined in the text. (RST.9-10.3)

Craft and Structure

- Determine the meaning of symbols, key terms, and other domain-specific words and phrases as they are used in a specific scientific or technical context relevant to grades 9-10 texts and topics. (RST.9-10.4)

Integration of Knowledge and Ideas

- Translate quantitative or technical information expressed in words in a text into visual form (e.g., a table or chart) and translate information expressed visually or mathematically (e.g., in an equation) into words. (RST.9-10.7)

- Compare and contrast findings presented in a text to those from other sources (including their own experiments), noting when the findings support or contradict previous explanations or accounts. (RST.9-10.9)

Range of Reading and Level of Text Complexity

- By the end of grade 10, read and comprehend science/technical texts in the grades 9-10 text complexity band independently and proficiently. (RST.9-10.10)

Writing

Research to Build and Present Knowledge

- Conduct short as well as more sustained research projects to answer a question (including a self-generated question) or solve a problem; narrow or broaden the inquiry when appropriate; synthesize multiple sources on the subject, demonstrating understanding of the subject under investigation. (W.9-10.7)

Investigating Bird Biodiversity Across the Americas

Before You Start...

Time

Preparation: **15 minutes**

Instruction: **45-60 minutes**

Preparation

- Make copies of the Student Reading and Worksheets.

- Download and make copies of the *Crossing Boundaries Excel Card* (optional) http://gisetc.com/bwb/

- Download the *"Birds 6 Data"* Excel spreadsheet and load it onto student workstations.

- Download the *"Birds 6 Teacher Slides"* PowerPoint file.

Technology

A-V Equipment

- LCD Projector
- Screen or Whiteboard

Computers

- Internet Connection

Information and Communication Technology

- Excel for creating bar graphs

Low-Tech Options

- Have students create graphs by hand.

Materials

- *"Birds 6 Data"* spreadsheet
- Colored pencils or markers (red, orange, yellow, green, and blue; one set per student or group)
- *Crossing Boundaries Excel Card* (optional)

Student Reading

- Biodiversity and Species Richness

Worksheets

- Student Reading Guides (Select from these options):
    - Anticipation Guide
    - Graphic Organizer
    - Reading Questions
    - Guided Reading
- Bird Species Richness

Investigation 6

Investigating Bird Biodiversity Across the Americas
The Investigation

Investigation Overview

1. Introduce the idea that biodiversity is not uniformly distributed across the planet.

2. Read and reflect on background information about biodiversity.

3. Using graphs, explore variation in bird species richness across North and South America.

Learning Objectives

Students will be able to:

- Use quantitative data to analyze spatial relationships and interpret differences in biodiversity across North and South America.

- Create bar graphs and assess trends in the number of bird species across countries.

- Analyze graphs and spatial data to explain latitudinal gradients in bird species richness across North and South America.

Key Concepts

- Biodiversity
- Species richness
- Population distribution
- Latitudianl gradient

Technology Overview

- With provided data, students use Excel to produce bar charts comparing the numbers of bird species by country in the Western Hemisphere.

Conducting the Investigation

1. Introduce the idea that biodiversity is not uniformly distributed across the planet.

Slide 2

   a. Project and discuss the Vertebrate Species Density map (Slide #2). Explain that this map shows variation in species density and ask the class what they think this term means.

   Species density is the number of species per unit of land area, such as # species/km².

   b. Ask: Where are the highest species densities found?

   Species densities are much higher in the tropics than near the poles. Some related facts:

   - *Tropical rainforests cover roughly 6% of the Earth's land surface but provide habitat to more than 50% of all species.*

   - *Of the roughly 10,000 bird species in the world, over 3,000 live in the Amazon River Basin. In comparison, fewer than 700 species of birds live in the continental United States even though the U.S. covers a larger area.*

2. Read and reflect on background information about biodiversity.

   a. Familiarize students with key concepts and vocabulary related to biodiversity by having them read the provided article entitled **Biodiversity and Species Richness** and completing one of the graphic organizers or guided reading sheets.

      - Option 1: Anticipation Guide Students answer True or False to a variety of statements before reading the article. After reading, they answer these same questions again and provide evidence from the reading to support each answer.

      - Option 2: Graphic Organizer Students read the article and then complete the graphic organizer based on this reading.

      - Option 3: Reading Questions (higher level option) Students read the article and then answer a series of questions to consolidate and demonstrate their understanding of key ideas and concepts.

      - Option 4: Guided Reading (lower level option) Students read the article and then complete a guided reading sheet to highlight key ideas and concepts.

      - Optional Extension for AP Classes: Students read the longer article titled "Biodiversity" by the Organization for Tropical Studies and write a summary of the article.

3. Using graphs, explore variation in bird species richness across North and South America.

   a. Begin by having students create a simple bar graph using the data in the "Part 1 Data" tab of the **Birds 6 Data** spreadsheet. This will provide experience in creating Excel graphs while also introducing the idea of species distribution. The Excel Card handout provides instructions for creating bar graphs in Excel 2010. If you have a different version of the software, some editing of these directions may be needed.

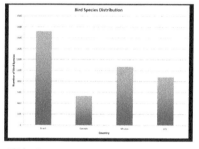

Slide 3

   b. After students have answered the questions on the **Bird Species Richness Worksheet**, discuss whether they can see any trends using these four data points that they could test using data for a larger number of countries. (Slide #3 shows this graph for projection.)

      If you rearrange the countries according to latitude, there appears to be a north/south trend with a progressively greater number of species in each country from Canada southward to Brazil. (Slide #4)

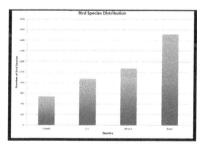

Slide 4

   c. Ask: Do you think this trend will hold true when additional countries are included? (Slide #5)

   d. Using data in the "Part 2 Data" tab, have students create a similar bar graph showing the number of bird species in a greater number of countries across the Western Hemisphere.

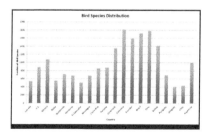

   e. Looking at this new graph (Slide #6), discuss whether the north/south trend holds true.

      No, at least not as cleanly as when only Canada, the U.S., Mexico, and Brazil are included.

Slide 6

f. Project the map of North and South America (Slide #7) and ask whether any trends are evident that relate to latitude.

> *The countries with highest numbers of bird species fall within 30°*
> *north or south of the Equator.*
> *Note: It may be worth noting that the alignment of countries along*
> *the x-axis involves some arbitrary decisions because the countries*
> *are not neatly lined up north and south of each other. Some of the*
> *long, narrow countries stretch across many lines of latitude and do*
> *not sit clearly north or south of others.)*

Slide 7

g. Wrap up by explaining that in the next Investigation we will take a more careful look at spatial distribution of bird species.

Biodiversity and Species Richness
Student Reading

What is biodiversity?

Biodiversity is a measure of the variety of species, genes, and ecosystems in a given area. Each *species* is made up of genetically related organisms that can successfully interbreed. One measure of biodiversity is *species richness*, the total number of species within a defined area.

About 1.8 million species have been described on Earth. Insects represent the most diverse taxonomic group, followed by plants. Vertebrates

Biodiversity also includes *genetic diversity*, the variety of genes within a species and among species. Genes contain DNA molecules that pass inherited biological traits from one generation to the next. Genetic diversity within a species shows up as differences among individuals of that species. Examples include the color of a human's eyes, or the size and shape of a dog. Species with high genetic diversity are more able to adapt to changes in their environment. When populations of a species become too small, genetic diversity is

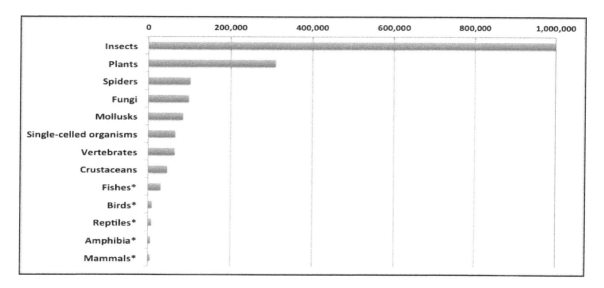

Figure 1. Numbers of described species in various taxonomic groups. Note that groups marked with * are vertebrates.

represent only 3% of Earth's biodiversity, and bird species make up only a fraction of vertebrates (Figure 1).

Many species have not yet been discovered, and scientists estimate that the true number of species on Earth could be over ten million. About 20,000 new species are identified each year. Most are in groups such as insects that have been less heavily studied than the vertebrates, but occasionally even new vertebrate species are found. As a result of the rapid rate of environmental change, many species likely are going extinct before they have even been discovered.

lost and the species becomes less resilient to change. Biodiversity loss is an important conservation issue, with increasing numbers of species becoming extinct due to habitat loss and environmental change.

The third measure of biodiversity is *ecosystem diversity*. An *ecosystem* is defined to be a community of species interacting with one another and with the nonliving environment. A grouping of similar ecosystems makes up an *ecoregion* whose plant and animal communities have adapted to similar conditions related to climate, soil, and other environmental factors. At the broadest scale, similar ecoregions can be lumped together into **biomes**. The climate and type of vegetation in each biome determine what other types of life it supports.

Major terrestrial biomes include forests, grasslands, deserts, and tundra. Tundra exists at the northernmost latitudes and also at lower latitudes on mountaintops where wintry weather is too intense for tree growth. Forest biomes

even millions of years. In any species, some organisms with genetic mutations survive while others do not. Mutations that enhance survival are more likely to be inherited and passed on to future generations. Over the course of many generations, enough genetic differences may accumulate that individuals of a given species cannot or will not interbreed. At this point, the single species splits into two.

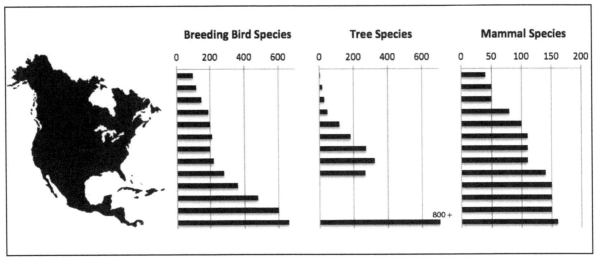

Figure 2. North-south gradient of bird, tree, and mammal species diversity. The numbers of species indicated in the bar graphs correspond to the latitude in the map at left. Tree species diversity is not available for some lower latitudes. (Adapted from Briggs, 1995, Global Biogeography, Amsterdam: Elsevier).

include boreal forests in Canada, temperate deciduous forests in the eastern United States, and temperate tropical forests in the northwestern United States. Near the Equator, both tropical rainforests and tropical deciduous forests exist. Grassland and desert biomes are found in midwestern and western United States.

The Amazon rainforest in Brazil represents one of the world's wettest biomes. Just to its west lies the world's driest hot desert. Stretching from Peru to Chile in a narrow strip between the Andes Mountains and the Pacific Coast, the Atacama Desert includes places where no rainfall has been recorded in the past 400 years!

How does biodiversity arise?

New species develop due to genetic mutations that lead to evolution. This involves many generations and takes place over thousands or

How is biodiversity distributed?

Biodiversity is not evenly distributed across the planet. On land, species diversity is lowest near the poles and increases toward the tropics (Figure 2). Tropical rainforests occupy only about 6% of Earth's land surface but may contain more than half the world's species. Much of this diversity is due to the high number of insect species, but tropical forests also provide habitat to more species of birds, mammals, and plants than most temperate regions. Almost a third of the world's bird species are found in the Amazon Basin.

Why do more species live near the equator than near the poles?

Scientists are trying to answer this question. One factor is that solar radiation hits the Earth more directly near the equator, supporting more photosynthesis and primary production. Another

Investigation 6

factor is that the climate is more stable near the equator, with greater consistency in temperature and precipitation throughout the year. Geological history also may help to explain the latitudinal gradient in species diversity. At northern latitudes, the landscape was repeatedly covered with glaciers that wiped out all plant and animal life while species in the tropics continued to thrive and evolve.

Why are some species more at risk than others?

Some species are adapted to living under a narrow set of habitat requirements. These species may have very small ranges. For example some bird species are found only on a single island in the Galapagos. Others live only within a narrow band of elevations in the Andes Mountains. Species such as these are called *endemic*, meaning that they are native to just one location (such as a single country or state, or even a single island). Endemic species face a high risk of extinction. Their limited distribution makes them sensitive to habitat loss, and their specialized habitat requirements place them at risk when faced with environmental change.

Biological "hotspots" are areas that host a relatively large number of specialized species that are found nowhere else in the world. Hotspots represent not only the richest but also the most threatened reservoirs of plant and animal life on Earth. The degree of threat is measured primarily in terms of habitat loss. The hotspots identified by Conservation International have lost at least 70 percent of their original natural vegetation. Insects, birds, and other forms of life that depend on these plant species are highly threatened.

This reading was adapted from a longer article by the Organization for Tropical Studies. The original article entitled "Biodiversity" includes additional information about the importance of biodiversity in providing ecosystem services and is available at http://www.ots.ac.cr/images/downloads/global-programs/resource-tools/biodiversity.pdf.

Investigation 6

Name: _____ Period: _____ Date: _____

Biodiversity & Species Richness
Anticipation Guide

Classify each statement as either *True/False* <u>BEFORE</u> you read the article. After reading the article, classify each statement again and then provide the evidence to back up your answer from the reading. Use the back of the paper if more space is needed.

BEFORE READING	AFTER READING	STATEMENT	EVIDENCE FROM THE ARTICLE
		Biodiversity can be measured in terms of species richness and genetic diversity.	
		There are approximately 500,000 species on Earth, with about 100 new organisms discovered each year.	
		Vertebrates are the taxonomic group with the largest number of species on Earth.	

Investigation 6

BEFORE READING	AFTER READING	STATEMENT	EVIDENCE FROM THE ARTICLE
		Mutations can be beneficial to organisms.	
		An endemic species lives in a variety of biomes around the world.	
		Biodiversity hotspots are areas of increasing biodiversity around the world.	
		Biodiversity is evenly distributed on Earth.	
		The distribution of species depends on the intensity of solar radiation and other factors.	

Investigation 6

Name: _____ Period: _____ Date: _____

Biodiversity & Species Richness
Anticipation Guide Answer Key

Classify each statement as either *True/False* <u>BEFORE</u> you read the article. After reading the article, classify each statement again and then provide the evidence to back up your answer from the reading. Use the back of the paper if more space is needed.

BEFORE READING	AFTER READING	STATEMENT	EVIDENCE FROM THE ARTICLE
		Biodiversity can be measured in terms of species richness and genetic diversity.	*TRUE* *Biodiversity includes the number of species but also the diversity of genes within each species. When a species is endangered, its genetic diversity drops and the species becomes less resilient to change.*
		There are approximately 500,000 species on Earth, with about 100 new organisms discovered each year.	*FALSE* *At least 1.7 million species have been described on Earth, and 20,000 new species are identified each year.*
		Vertebrates are the taxonomic group with the largest number of species on Earth.	*FALSE* *Insects are the most diverse group.*

Investigation 6

BEFORE READING	AFTER READING	STATEMENT	EVIDENCE FROM THE ARTICLE
		Mutations can be beneficial to organisms.	*TRUE* *Some mutations enhance survival, and these are more likely to be inherited and passed on to future generations.*
		An endemic species lives in a variety of biomes around the world.	*FALSE* *An endemic species lives only within a single defined region such as a country or state.*
		Biodiversity hotspots are areas of increasing biodiversity around the world.	*FALSE* *Biodiversity hotspots are rich in biodiversity, but they also are the areas in which this diversity is most highly threatened.*
		Biodiversity is evenly distributed on Earth.	*FALSE* *Species diversity is lowest near the poles and increases toward the tropics.*
		The distribution of species depends on the intensity of solar radiation and other factors.	*TRUE* *Other factors include climate and geological history such as presence vs. absence of glaciers over geologic time.*

Biodiversity & Species Richness
Graphic Organizer

Biodiversity is:
(in your own words)

Species means

Species Richness is a measure of biodiversity because

Ecosystems are

Biomes are

Ecoregions are different based on

Endemic species are

Biological Hotspots are special because

In your own words, SUMMARIZE how biodiversity is distributed around the world using information contained in Figure 2 and using at least 4 of the vocabulary words above.

Investigation 6

Name: _____ Period: _____ Date: _____

Biodiversity & Species Richness
Graphic Organizer Answer Key

Biodiversity is:

(in your own words)

the diversity or variety of life on Earth including the number of species and the genetic variation with each species.

Species means

a collection of organisms that are genetically related and can interbreed.

Species Richness is a measure of biodiversity because

it describes the number of species in area and this is a key aspect of biodiversity.

Ecosystems are

communities of species that interact with each other and with the environment.

Biomes are

regions characterized by their climate and type of vegetation.

Ecoregions are different based on

plant and animal communities that have adapted to specific environmental conditions.

Endemic species are

species that are native to and live only within a defined region such as a country or state.

Biological Hotspots are special because

they host many endemic species.

In your own words, SUMMARIZE how biodiversity is distributed around the world using information contained in Figure 2 and using at least 4 of the vocabulary words above.

Biodiversity is highest at the Equator. The US has fewer species of birds, trees and mammals than in the tropics. Tropical rain forests' biomes have the highest biodiversity on Earth. Biological hotspots contain many endemic species, including some that are highly threatened.

Investigation 6

Biodiversity & Species Richness
Reading Questions

1. Define, in your own words, the term "biodiversity."

2. What is a "species," and what does "species richness" mean?

   Species:

   Species Richness:

3. Why do scientists use species richness as a measure of biodiversity?

4. Describe the main difference between an ecosystem and a biome.

5. Describe a location in the world where the following biomes could be found:

   Desert:

   Tundra:

   Tropical Forest:

   Temperate Grassland:

6. How many species have been identified by biologists?

7. How many new species are identified each year?

8. Using Figure 1, which group makes up the largest percentage of organisms on Earth, and approximately what percentage does his group represent? (Give the approximate percentage from the graph.)

9. Which group is second, and at what percentage?

10. Name three groups of organisms that are vertebrates.

    a)

    b)

    c)

11. Of all species on Earth, what percentage are vertebrates?

12. Why is it important for a species to have a high genetic diversity?

13. Evolution can lead to the development of new species over time. What is it that leads to variation at the genetic level?

14. How does the variation mentioned in Question 11 lead to formation of a new species?

15. Define an "endemic species."

16. Why do endemic species have high rates of extinction?

17. Describe a biological hotspot and explain the importance of protecting hotspots around the world.

18. Is biodiversity distributed evenly across the planet? Use Figure 2 to help support your answer with data.

19. Which terrestrial biome on Earth has the highest biodiversity? Support your answer.

20. Why does such variation in biodiversity occur around the world?

Name: _____ Period: _____ Date: _____

Biodiversity & Species Richness
Reading Questions Answer Key

1. Define, in your own words, the term "biodiversity."

   The diversity or variety of life on Earth, including the number of species and the genetic variation within each species.

2. What is a "species," and what does "species richness" mean?

   species: *a collection of organisms that are genetically related and can interbreed*

   species richness: *the number of species within an area of interest*

3. Why do scientists use species richness as a measure of biodiversity?

   Species richness describes the number of species in an area, and this is a key aspect of biodiversity.

4. Describe the main difference between an ecosystem and a biome.

   An ecosystem is a community of species interacting with each other and with the environment. A biome includes more than one type of ecosystem and is a region that is characterized by its climate and type of vegetation.

5. Describe a location in the world where the following biomes could be found:

   Desert: *The article mentions deserts in midwestern and western parts of the U.S.*

   Tundra: *At northern latitudes and on mountaintops where weather is too intense for tree growth.*

   Tropical forest: *Temperate tropical forests in the northwestern US, and both tropical rainforests and tropical deciduous forests near the Equator.*

   Temperate grassland: *In Midwestern and Western parts of the U.S.*

6. How many species have been identified by biologists?

   About 1.7 million species

7. How many new species are identified each year?

   About 20,000

8. Using Figure 1, which group makes up the largest percentage of organisms on Earth, and approximately what percentage does his group represent? (Give the approximate percentage from the graph.)

    Insects make up about 58% of all described species on Earth.

9. Which group is second, and at what percentage?

    Plants, at about 18%

10. Name three groups of organisms that are vertebrates.

    Fish, birds, mammals, reptiles, and amphibians

11. Of all species on Earth, what percentage are vertebrates?

    3%

12. Why is it important for a species to have a high genetic diversity?

    Genetic diversity produces differences among individuals of a species. Species with high genetic diversity are more able to adapt to changes in their environment and therefore more less at risk of extinction.

13. Evolution can lead to the development of new species over time. What is it that leads to variation at the genetic level?

    Genetic mutations

14. How does the variation mentioned in Question 11 lead to formation of a new species?

    Some organisms with genetic mutations survive and pass their genes along to future generations. Over many generations, if genetic differences prevent interbreeding then a single species can split into two.

15. Define an "endemic species."

    An endemic species is one that is native to and lives only within a defined region such as a country, state, or smaller area.

16. Why do endemic species have high rates of extinction?

    Because they tend to be specialized rather than generalized and therefore cannot readily adapt to environmental change.

17. Describe a biological hotspot and explain the importance of protecting hotspots around the world.

    A hotspot is an area that hosts a large number of endemic species. It is important to protect them because they support these species and are threatened regions with a high rate of habitat loss.

18. Is biodiversity distributed evenly across the planet? Use Figure 2 to help support your answer with data.

    Biodiversity is not distributed evenly. Figure 2 shows that the numbers of species of birds, trees, and mammals increases from northern latitudes going south toward the Equator.

19. Which terrestrial biome on Earth has the highest biodiversity? Support your answer.

    Tropical rainforests support the highest biodiversity. They contain more than half of all species on Earth but occupy only about 6% of the land surface.

20. Why does such variation in biodiversity occur around the world?

    Scientists are still trying to answer this question, but factors include the intensity of solar radiation, stability of the climate in each region, and geologic history including history of glaciation.

Biodiversity & Species Richness
Guided Reading

1. Biodiversity refers to _____.

2. A _____ includes genetically related organisms that can successfully interbreed.

3. A biome is a region classified by its _____ and _____.

4. A community of species interacting with each other and with the nonliving environment would be considered an _____.

5. Major land biomes include _____, _____, _____, and _____.

6. _____ is the biome found at the northernmost latitudes. _____ is in the eastern United States. The _____ biome is found in the Midwestern United States. Near the _____ is where the tropical rainforest can be found.

7. Approximately _____ species have been identified on Earth, with about _____ new ones being identified each year.

8. The most diverse group of organisms is _____ at approximately _____%. Plants are the second most abundant groups at approximately _____%. _____, which include fish, birds, mammals, reptiles, and amphibians, make up only _____%.

9. In addition to species diversity, biodiversity also includes _____ diversity, which is the variety of _____ within a species and among species. Examples of genetic diversity include _____ and _____.

10. Species with high genetic diversity are more resilient because they are better able to _____.

11. Biodiversity arises due to _____ that lead to evolution. Those that enhance survival are more likely to be _____ and passed on to _____.

12. _____ live only within one defined region and are native to that region. For birds, endemism commonly refers to where they _____. A bird species could be

endemic to a single _____, _____, or _____.

13. Endemic species tend to be highly specialized. They are therefore not able to adapt to

_____. This causes them to face a higher risk of

_____.

14. Biological "hotspots" are areas that host large numbers of

_____. They are typically defined by the number of

_____ species. In addition to being the richest reservoirs of plant and animal life,

hotspots are also the most _____ reservoirs. The degree of threat is

measured mainly in terms of _____.

15. Biodiversity is not evenly distributed. It is lowest near the _____ and

_____ towards the tropics. Tropical rainforests only occupy about _____%

of the land surface but may have more than half of the species on Earth. Almost _____ of

the world's bird species are found in the Amazon.

16. Three factors that contribute to there being more species living near the Equator versus the poles

are:

a) _____

b) _____

c) _____

Name: _____ Period: _____ Date: _____

Biodiversity & Species Richness
Guided Reading Answer Key

Investigation 6

1. Biodiversity refers to ___*the variety of life on Earth*___.

2. A ___*species*___ includes genetically related organisms that can successfully interbreed.

3. A biome is a region classified by its ___*climate*___ and ___*vegetation*___.

4. A community of species interacting with each other and with the nonliving environment would be considered an ___*ecosystem*___.

5. Major land biomes include ___*forests*___, ___*grasslands*___, ___*desert*___, and ___*tundra*___.

6. ___*Tundra*___ is the biome found at the northernmost latitudes.

   ___*Forest (temperate deciduous forest)*___ is in the Eastern United States. The ___*grassland*___ biome is found in the Midwestern United States. Near the ___ ___*Equator*___ is where the tropical rainforest can be found.

7. Approximately ___*1.7 million*___ species have been identified on Earth, with about ___*20,000*___ new ones being identified each year.

8. The most diverse group of organisms is ___*insects*___ at approximately ___*58*___%. Plants are the second most abundant groups at approximately ___*18*___%. ___*Vertebrates*___, which include fish, birds, mammals, reptiles, and amphibians, make up only ___*3*___%.

9. In addition to species diversity, biodiversity also includes ___*genetic*___ diversity, which is the variety of ___*genes*___ within a species and among species. Examples of genetic diversity include ___*the color of a human's eyes*___ and ___*the size and shape of a dog*___.

10. Species with high genetic diversity are more resilient because they are better able to ___*adapt to changes in the environment*___.

11. Biodiversity arises due to ___*genetic mutations*___ that lead to evolution. Those that enhance survival are more likely to be ___*inherited*___ and passed on to ___*future generations*___.

12. ___*Endemic species*___ live only within one defined region and are native to that region. For

birds, endemism commonly refers to where they ___breed___. A bird species could be

endemic to a single ___state___, ___ecoregion___ or ___country___.

13. Endemic species tend to be highly specialized. They are therefore not able to adapt to

    ___environmental change___. This causes them to face a higher risk of ___extinction___.

14. Biological "hotspots" are areas that host large numbers of ___endemic species___.
    They are typically defined by the number of ___plant___ species. In addition to being the richest
    reservoirs of plant and animal life, hotspots are also the most ___threatened___ reservoirs. The
    degree of threat is measured mainly in terms of ___habitat loss___.

15. Biodiversity is not evenly distributed. It is lowest near the ___poles___ and ___higher___ towards
    the tropics. Tropical rainforests only occupy about ___6___% of the land surface but may have more
    than half of the species on Earth. Almost ___a third___ of the world's bird species are found in the
    Amazon.

16. Three factors that contribute to there being more species living near the Equator versus the poles
    are:

    a) *Solar radiation hits the Earth more directly near the Equator, supporting more photosynthesis and primary production.*

    b) *The climate is more stable near the Equator.*

    c) *Geologic history – plant and animal life in the north were wiped out by glaciers but this did not occur in the tropics.*

Bird Species Richness
Worksheet

Part 1

1. Open the *"Bird Investigation 6 Data"* spreadsheet and select the tab labeled *"Part 1 Data."* Using the Charts function in Excel, create a bar graph showing the number of bird species in Brazil, Canada, Mexico, and the U.S.

2. Compare your graph with the sample graph shown here. If they are not identical, review your procedures and make any needed corrections.

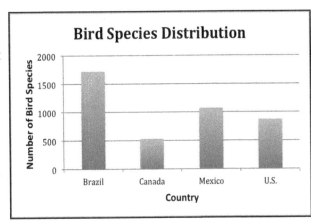

Bird Species Distribution

3. Use the graph to answer these questions:

   a. Which of these countries has the greatest number of bird species?

   b. Why is a bar graph used to represent these data instead of a line graph?

   c. The countries along the x-axis are arranged in alphabetical order. What other arrangement would make more sense geographically? Sketch that graph here:

   d. Now that you have arranged the countries geographically, what trend becomes apparent?

Part 2

1. Using the tab labeled "*Part 2 Data*," create a bar chart to represent the number of bird species in selected countries across North and South America. As before, place the countries on the x-axis and the number of bird species on the y-axis. Be sure to label the axes, make a key, and add a title to your graph.

2. Describe any trends that you observe:

Bird Species Richness
Worksheet Answer Key

Part 1

1. Open the *"Bird Investigation 6 Data"* spreadsheet and select the tab labeled *"Part 1 Data."* Using the Charts function in Excel, create a bar graph showing the number of bird species in Brazil, Canada, Mexico, and the U.S.

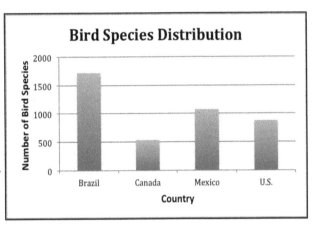

2. Compare your graph with the sample graph shown here. If they are not identical, review your procedures and make any needed corrections.

3. Use the graph to answer these questions:

   a. Which of these countries has the greatest number of bird species?

      Brazil

   b. Why is a bar graph used to represent these data instead of a line graph?

      Each country has a distinct number of bird species, represented by a bar. If we use a line graph instead, that would imply a connection linking the countries.

   c. The countries along the x-axis are arranged in alphabetical order. What other arrangement would make more sense geographically? Sketch that graph here:

      It makes more sense intuitively to arrange them in geographic order. Here's what it looks like when the countries are arranged from north to south.

      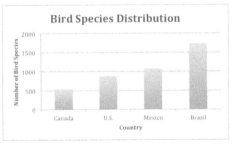

   d. Now that you have arranged the countries geographically, what trend becomes apparent?

      There appears to be a trend showing increasing numbers of species in countries closer to the Equator.

Investigation 6

Part 2

1. Using the tab labeled "*Part 2 Data*," create a bar chart to represent the number of bird species in selected countries across North and South America. As before, place the countries on the x-axis and the number of bird species on the y-axis. Be sure to label the axes, make a key, and add a title to your graph.

2. Describe any trends that you observe:

   The greatest numbers of bird species are found in countries on or near the Equator: Columbia, Peru, Brazil, and Ecuador. There's a dip in Central America and another dip to the south in Paraguay, Uruguay, and Chile.

Investigation 6

7

investigation

Exploring Bird Conservation Needs at Home and Abroad

Investigation 7 Photo Credits

Phil Kahler. Snowy Owl. (https://flic.kr/p/dNPF1B)

Phil Kahler. Long Billed Dowitcher. (https://flic.kr/p/dRf59p)

Phil Kahler. Black Turnstone. (https://flic.kr/p/oXzCzk)

Investigation 7
Exploring Bird Conservation Needs at Home and Abroad

Using graphs and maps, students explore the spatial distribution of bird species across the Western Hemisphere. They investigate factors affecting species distribution, consider the implications of annual life cycles of migratory birds, and reflect on the implications for conservation.

Essential Questions

- What are some reasons for the variation in distribution of bird species across North and South America?

- What countries or regions should be targeted for conservation of bird species?

- Why does conservation of some migratory species require international collaboration?

Science Concepts and Topics

Ecologists have identified **biological hotspots**, areas that host relatively large numbers of **endemic species** found nowhere else in the world. Such regions are priority targets for conservation. However, conservation efforts also must encompass efforts to protect the more widely distributed species to ensure that common birds remain common and that threatened species will not be driven to extinction.

In the bird world, many species migrate across continents. These species rely on high quality habitat in different locations at each stage of their annual life cycle. For example, reproductive success depends on availability of food and nesting sites during the breeding season. The health and survival of individuals who will breed in future years depends on the quality of habitat where they live during the winter months. Many species also depend on stopover sites where they can rest and feed in the midst of long migratory journeys. Full life-cycle protection of migratory species may depend on multinational cooperation because protection of habitats used during any one season will be insufficient if those used at other times of year are threatened.

Standards Addressed in This Investigation

Next Generation Science Standards – Middle School

- Analyze and interpret data to provide evidence for the effects of resource availability on organisms and populations of organisms in an ecosystem. (MS-LS2-1)

- Construct an explanation that predicts patterns of interactions among organisms across multiple ecosystems. (MS-LS2-2)

Next Generation Science Standards – High School

- Use mathematical representations to support and revise explanations based on evidence about factors affecting biodiversity and populations in ecosystems of different scales. (HS-LS2-2)

- Evaluate the evidence for the role of group behavior on individual and species' chances to survive and reproduce (HS-LS2-8)

Investigation 7

Next Generation Science Standards – Scientific Practices

- Asking questions and defining problems
- Planning and carrying out investigations
- Analyzing and interpreting data
- Using mathematics and computational thinking
- Constructing explanations and designing solutions
- Engaging in argument from evidence
- Obtaining, evaluating, and communicating information

Common Core (9-10th grade)

Reading Science & Technical Subjects

Key Ideas and Details

- Follow precisely a complex multistep procedure when carrying out experiments, taking measurements, or performing technical tasks, attending to special cases or exceptions defined in the text. (RST.9-10.3)

Craft and Structure

- Determine the meaning of symbols, key terms, and other domain-specific words and phrases as they are used in a specific scientific or technical context relevant to grades 9-10 texts and topics. (RST.9-10.4)

Integration of Knowledge and Ideas

- Translate quantitative or technical information expressed in words in a text into visual form (e.g., a table or chart) and translate information expressed visually or mathematically (e.g., in an equation) into words. (RST.9-10.7)

Writing

Research to Build and Present Knowledge

- Conduct short as well as more sustained research projects to answer a question (including a self-generated question) or solve a problem; narrow or broaden the inquiry when appropriate; synthesize multiple sources on the subject, demonstrating understanding of the subject under investigation. (W.9-10.7)
- Gather relevant information from multiple authoritative print and digital sources, using advanced searches effectively; assess the usefulness of each source in answering the research question; integrate information into the text selectively to maintain the flow of ideas, avoiding plagiarism and following a standard format for citation. (W.9-10.8)

Speaking and Listening

Comprehension and Collaboration

- Initiate and participate effectively in a range of collaborative discussions (one-on-one, in groups, and teacher-led) with diverse partners on grades 9-10 topics, texts, and issues, building on others' ideas and expressing their own clearly and persuasively. (SL.9-10.1)
- Integrate multiple sources of information presented in diverse media or formats (e.g., visually, quantitatively, orally) evaluating the credibility and accuracy of each source. (SL.9-10.2)

Exploring Bird Conservation Needs at Home and Abroad

Before You Start...

Time

Preparation: **15 minutes**

Instruction: **90 minutes**

Preparation

- Make copies of the Worksheets and the Web Map Help Sheet.

- Download the *"Birds 7 Data"* Excel spreadsheet and load it onto student workstations.

- Download the *"Birds 7 Teacher Slides"* PowerPoint file.

- Access the *"Birds 7 ArcGIS Map"*
  http://crossingboundaries.org/bwb.php

- Follow these steps only if you will be using the Interactive PDF instead of ArcGIS Online:

  - Install Acrobat Reader from www.get.adobe.com/reader, on student computers.

  - Download the *"Birds 7 Map"* Interactive PDF.

  - Make this interactive PDF accessible on student computers and test the functionality of the layers.

Technology

A-V Equipment

- LCD Projector
- Screen or Whiteboard

Computers

- Web access for ArcGIS Online

If using Interactive PDF option:

- Acrobat Reader
- *"Birds 7 Map"* Interactive PDF

Information and Communication Technology

- Excel for creating bar graphs
- ArcGIS Online

Low-Tech Options

- Use the Interactive PDF map. If student computers are unavailable, print and/or photocopy the layers of the *"Birds 7 Map"* Interactive PDF for use as transparent overlays or handouts.

Materials

- *"Birds 7 Map"* (online or interactive PDF)
- *"Birds 7 Data"* spreadsheet
- Web Map Help Sheet
- Colored pencils or markers (red, orange, yellow, green, and blue; one set per student or group)

Worksheets

- Bird Species Richness across North and South America
- Distribution of Bird Species Richness
- Globally Threatened Bird Species

Exploring Bird Conservation Needs at Home and Abroad
The Investigation

Investigation Overview

1. Using maps, explore the variation in bird species richness across latitudinal gradients in North and South America.

2. Consider why some bird species migrate.

3. Compare bird species richness with geographic and ecological factors.

4. Explore distribution of globally threatened bird species.

5. Think about how to set bird conservation priorities.

Learning Objectives

Students will be able to:

- Use maps and quantitative data to explore the variation in bird species richness across latitudinal gradients in North and South America.

- Evaluate the pros and cons of using political versus ecological boundaries in analyzing bird biodiversity.

- Explain the concept of full life-cycle conservation and why it is important for migratory birds.

Key Concepts

- Biodiversity
- Species richness
- Latitudinal gradient
- Ecoregion
- Biological hotspot
- Life-cycle conservation of migratory birds

Technology Overview

- Students explore geographic and ecological aspects of bird species richness using ArcGIS Online or map layers in an interactive PDF, along with bar charts produced in Excel.

Conducting the Investigation

1. Using maps, explore the variation in bird species richness across latitudinal gradients in North and South America.

   a. Project Slide #2 showing the number of bird species per country and ask the students to color-code the **Bird Species per Country** map in the **Bird Species Richness across North and South America Worksheet** (Slide #3) and complete the key with their selected colors. (For consistency with the Vertebrate Species Density figure shown in Investigation 6, you may want to use red for the greatest number of species, then orange, yellow, green, and blue for the lower categories.)

   b. Using evidence from the students' maps, continue the discussion begun in Investigation 6 about trends in bird species distribution across North

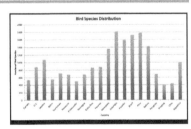

Slide 2

Slide 3

Investigation 7

and South America. Ask what factors other than latitude are likely to affect distribution of bird species.

- *Country size – One factor is the overall size of each country. You might expect larger countries to host more species than smaller ones. However, there are exceptions. For example, Peru is roughly 1/8th the size of Canada but has over three times as many bird species.*

- *Migratory behavior – Some species migrate and others do not. Therefore, some species are widely distributed across many countries while others are not.*

c. Display and discuss the *Density of Bird Species* graph (Slide #4). What countries show the highest peaks? Why might this be the case?

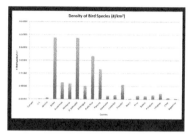
Slide 4

Several tiny countries have the highest bird densities – Belize and El Salvador, followed by Costa Rica and Panama. Factors include:

- *Habitat diversity – Although these countries are small, they contain a wide variety of habitats. These include rainforests and cloud forests, rocky shores and sandy beaches, grassy plains and steep mountains, lakes and wetlands. Collectively these habitats provide food and shelter for many species of birds.*

Slide 5

- *Biome – Recall from the reading that tropical rainforests are richer in species than temperate forests or other higher-latitude biomes. Use a biome map to view the extent of the "Tropical and Subtropical Moist Broadleaf Forests" biome (Slide #5).*

2. Consider why some bird species migrate.

a. Investigation 3 focused on the amazing phenomenon of bird migration. Now let's look at how migration varies from country to country across the Western Hemisphere. Over half of the world's roughly 10,000 species of birds are migratory, but the percentage varies greatly with latitude. Either have students create a graph showing percentages of bird species that are migratory, or display the graph provided (Slide #6). If they will be creating their own graphs, they will use data in the *"Migratory Bird Data"* tab of the **Birds 7 Data** spreadsheet.

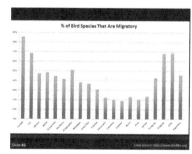
Slide 6

b. Discuss what trend is evident in the migration graph, and what this indicates in terms of bird life cycles.

All countries have some non-migratory species that stay in place year-round and others that migrate. In general, tropical countries have lower percentages of migratory species compared with countries farther from the Equator. In Canada and the U.S., many species escape the harsh winters by flying south. We think of them as "our birds," but of course another way of looking at these migratory species is as tropical residents that fly to our part of the world to breed because space is abundant and there is less competition for high-protein food to feed their young.

3. Compare bird species richness with geographic and ecological factors.

a. Project the *Bird Species Richness Map* (Slide #7) to introduce the idea that students will be relating the distribution of species richness to various environmental variables.

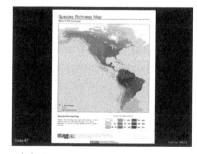
Slide 7

b. Working individually or in small groups, have students complete the **Distribution of Bird Species Richness Worksheet** using the ArcGIS Online *Birds 7 Map* or interactive PDF file to research the climate and terrain of the areas with the highest concentrations of bird species. The goal is to explore variables such as climate, terrain, precipitation, and land cover to see how these environmental factors relate to the distribution of bird species richness.

c. Ask students to use the worksheet to record and analyze their findings.

d. Discuss the worksheet as a class. The following factors help to explain the patterns in the data:

- *Temperature* – *In general, regions with consistently warm climates support more species than those with temperature extremes.*

- *Water* – *Water is another key factor. More species live in regions where moisture is available year-round than in those with periodic drought.*

- *Habitat diversity* – *The greater the variability in habitat, the greater number of species likely to be supported in a region.*

e. Project the GIS or Interactive PDF file and turn on the "Countries" layer. Compare country borders with natural features including biomes, ecoregions, and topographic features.

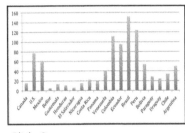

Slide 8

f. Discuss the advantages vs. disadvantages of planning bird conservation measures on a country level.

Country boundaries are useful because of the ability of national governments to pass and enforce regulations and to create protected areas within their borders. However, birds follow ecological rather than political boundaries, so describing bird distribution by country can be misleading and protection efforts are incomplete if they stop at country borders.

4. Explore distribution of globally threatened bird species.

a. In the *Birds 7 Data* spreadsheet, have students select the tab labeled "*Globally Threatened Bird Data*" and use the data there to create a bar graph illustrating one of these categories:

- # Globally Threatened bird species per country

- % bird species that are Globally Threatened

- Density of Globally Threatened bird species (# species/km^2).

(Note: these graphs are provided as Slides 8-10 in case you prefer not to engage students in this additional graphing exercise, or for display and discussion after students have created their own graphs.)

b. Ask students to use the **Globally Threatened Bird Species Worksheet** to record and analyze their findings.

c. Ask each group to pair up with another that has analyzed a different category to compare results and determine what relationships they have discovered.

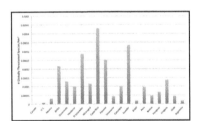

Slide 9

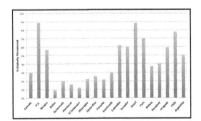

Slide 10

d. Discuss the students' findings, asking what differences they discovered in country rankings according to these three metrics.

# of Globally Threatened bird Species – *Brazil, Peru, and Columbia have the highest numbers. Looking at the map, you can see that these are relatively large countries on the Equator that include the Amazon River Basin. Of the roughly 10,000 bird species in the world, over 3,000 live in the Amazon. (In comparison, fewer than 700 bird species live in the entire continental U.S.)*

Effect of Land Area – *Costa Rica, Ecuador, and El Salvador come out on top when land area is taken into account. These relatively tiny Central American countries provide critical habitat for both resident and migratory species, including many threatened species. In addition to providing year-round and seasonal habitats, Central America also serves as a crucial stopover site for migratory species that pause there in the midst of traveling long distances between their winter homes in South America and breeding grounds to the north.*

Percentage of Total Bird Species – *U.S., Brazil, and Chile are top in Globally Threatened species as a percentage of total bird species. Although the U.S. hosts relatively few bird species per land area, a disproportionate number of our bird species are threatened. Every type of habitat in the U.S. hosts species in need of conservation action. This need is particularly acute in Hawaii, where many of the native bird species are endangered.*

5. Think about how to set bird conservation priorities.

a. Collectively discuss the ways in which you have viewed and portrayed various aspects of bird species richness:

- What are some ways in which the distribution of bird species varies across North and South America?

  Bird species richness is highest near the Equator and tends to decrease with distance toward the north and south poles.

- What factors do you think are most important in determining these distributions?

  Climate, geography, geologic history (lack of glaciation in tropical areas)

- What other data or information might we need to continue our investigation?

  Ecoregion characteristics, geologic features, weather and climate

Slide 11

b. Project Slide #11 and discuss the full life cycle needs of a migratory bird in terms of habitat in its breeding area, wintering zone, and possibly also stopover sites that provide places to rest and feed in the midst of long migratory flights.

c. Consider the value of political versus ecological boundaries in determining priories for conservation initiatives. Ask: Should we make plans for protection of endangered species based on political boundaries or using natural boundaries such as ecoregions or biomes? (Slide 11). Discussion points:

- Why are political boundaries useful in terms of conservation?

  Most laws, regulations, and conservation incentives are established and enforced by agencies that govern a particular country, state, or other unit.

- If we are making decisions based only on political boundaries, what challenges or disadvantages does that create?

  Birds don't follow political boundaries in choosing where to live, and migratory species are likely to depend on habitats in several different countries for their breeding sites, wintering areas, and stopover sites during migration.

- What are some of the ways governments might be able to do both?

  Establish conservation programs within their own borders while also collaborating with other governments and organizations to ensure protection of threatened species throughout their entire range and across all seasons.

d. Wrap up the Investigation by reflecting on the variation in bird biodiversity across the continent and the complexities of planning conservation programs that span political boundaries. This leads into the final Investigation, in which students select topics and develop action plans to protect or restore a selected bird species or critical habitat of multinational concern.

Web Map Help Sheet

Bird Species Richness—North and South America

A guide to navigation and use of the web map for this investigation.

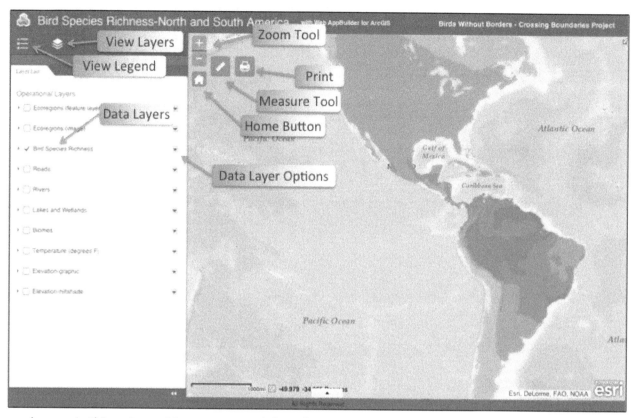

The web maps in this unit provide access to real data about species distributions and environmental conditions, landforms, and other geographical features. Each map includes a series of "data layers." You can view one layer at a time or turn on several simultaneously to explore how they relate.

When you first open a web map, the list of available data layers may be hidden. To make it visible, simply click on the View Layers icon. When a layer is turned on, its Legend becomes active. You can toggle back and forth between viewing the layer list and legends simply by clicking on the View Layers and View Legend icons.

The Data Layer Options symbol brings up a menu offering options including "Transparency." Try moving the Transparency slider and explore how that helps you compare two or more data layers.

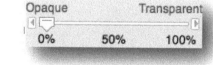

Icons in the top left corner of the map show available tools. All maps include tools for zooming in and out, measuring features, and printing your map. The Home button will bring the map back to the original perspective seen when it was first opened.

The best way to learn how to use one of these maps is to play with it. You can't break it!

Bird Species Richness Across North and South America Worksheet

1. On the *Bird Species per Country* map, illustrate the distribution of bird species richness in North and South America. Select five colors to represent the ranges shown in the map key, thinking about which will intuitively show high versus low values. Add these colors to the key and then color in the countries using data from your bar chart or the spreadsheet.

2. Answer the following questions, based on your map and knowledge of biodiversity:

   a. Which continent supports the largest number of bird species?

   b. What is "species richness"?

   c. In what range of latitudes do you find the countries with the highest bird species richness?

   d. Why do you think this region supports the greatest diversity in bird species?

Name: _____ Period: _____ Date: _____

Bird Species per Country Map

Image Source: NASA.gov

KEY	# Species	< 500	500-899	900-1299	1300-1699	> 1700
	Color on Map					

Bird Species Richness Across North and South America
Worksheet Answer Key

CROSSING
boundaries

1. On the **Bird Species per Country** map, illustrate the distribution of bird species richness in North and South America. Select five colors to represent the ranges shown in the map key, thinking about which will intuitively show high versus low values. Add these colors to the key and then color in the countries using data from your bar chart or the spreadsheet.

    Answers vary

2. Answer the following questions, based on your map and knowledge of biodiversity:

    a. Which continent supports the largest number of bird species?

      South America

    b. What is "species richness"?

      The number of species within an area

    c. In what range of latitudes do you find the countries with the highest bird species richness?

      15°N to 15°S or 30°S

    d. Why do you think this region supports the greatest diversity in bird species?

      This region is close to the Equator and contains tropical rainforests and other rich tropical habitats.

Distribution of Bird Species Richness
Worksheet

1. Using the ArcGIS Online **Birds 7 Map**, explore relationships between bird species richness and environmental variables. You will need to toggle on/off layers showing variables such as climate, elevation, annual precipitation, and land cover.

   You can use the slider to make the bird layer more or less transparent In order to investigate the climate and terrain of the areas with the highest concentrations of bird species. Your goal is to explore whether any of the environmental factors correlate well with the distribution of high bird species richness.

2. Summarize your findings in this table for three countries that have particularly high concentrations of bird species:

Country	Terrain	Landscape Features	Temperature	Precipitation	Other

3. In a few sentences, describe the terrain and climate for each of the three countries that you selected. Are there any similarities? Differences?

   a.

   b.

   c.

Investigation 7

Globally Threatened Bird Species
Worksheet

One out of every eight bird species in the world is threatened with extinction. How are these globally threatened species distributed across North and South America?

1. Using the tab labeled "Worksheet 4 Data," create a bar chart to show one of these three categories. Check the box in front of the category you are investigating, and add the appropriate title to your bar chart:

   - Number of Globally Threatened bird species per country

   - Percent of Bird Species that are Globally Threatened (%)

   - Density of Globally Threatened bird species ($\#/km^2$)

2. Which 3 countries in your graph have the highest numbers or proportions of globally threatened bird species?

   a._____

   b._____

   c._____

3. Highlight these three countries on the Globally Threatened Bird Species map.

4. Meet with a student or team that has explored a different category and compare your findings. Did the same countries rise to the top? What might explain any differences you have found?

Globally Threatened Bird Species Map

Canada

United States

Mexico

Belize

Nicaragua

Guatemala
Honduras
El Salvador
Costa Rica
Colombia
Ecuador

Panama

Venezuela

Peru

Brazil

Bolivia

Chile

Argentina

Paraguay

Uruguay

30° N

15° N

0° N

15° S

30° S

Image Source: NASA.gov

135°   120°   105°   90°   75°

KEY	Category					
	Color on Map					

Globally Threatened Bird Species
Worksheet Answer Key

One out of every eight bird species in the world is threatened with extinction. How are these globally threatened species distributed across North and South America?

1. Using the tab labeled "Worksheet 4 Data," create a bar chart to show one of these three categories. Check the box in front of the category you are investigating, and add the appropriate title to your bar chart:

    - Number of Globally Threatened bird species per country

    - Percent of Bird Species that are Globally Threatened (%)

    - Density of Globally Threatened bird species ($\#/km^2$)

2. Which 3 countries in your graph have the highest numbers or proportions of globally threatened bird species?

    Answers depend on category selected in #1.

# Globally Threatened bird species per country	% Bird Species that are Globally Threatened	Density of Globally Threatened bird species
Brazil	United States	Costa Rica
Peru	Brazil	Ecuador
Colombia	Chile	El Salvador

3. Highlight these three countries on the Globally Threatened Bird Species map.

4. Meet with a student or team that has explored a different category and compare your findings. Did the same countries rise to the top? What might explain any differences you have found?

8
investigation
Creating a Conservation Plan

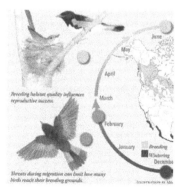

Investigation 8 Photo Credits

Phil Kahler. Spotted Owl. Personal Collection.

Phil Kahler. Pileated Woodpecker. (https://flic.kr/p/nDtr1Z)

Phil Kahler. Western Meadowlark. (https://flic.kr/p/e87Jwz)

Investigation 8
Creating a Conservation Plan

In this final Investigation, students select one of six conservation action pathways and then create a plan for protection or restoration of a specific bird species or critical habitat of concern in the U.S. and our neighboring countries.

Essential Questions

- What threats do birds face, in the U.S. and beyond our borders?
- What can be done to protect or restore one or more critical habitats or threatened species?

Science Concepts and Topics

The U.S., Canada, and Mexico are home to over 1150 species of birds, 148 of which are in need of immediate conservation attention because of their highly threatened and declining populations. Forty-two common bird species have steeply declined by 50% or more in the past 40 years. Habitat loss is the most common reason, but the good news is that conservation works. Several species near extinction have rebounded, and others can similarly succeed if given the chance.

Partners in Flight is a collaborative group representing bird conservation experts in Canada, the U.S., and Mexico. They aim to prevent extinction of bird species at greatest risk, ensure that common birds remain common, and promote the diversity and abundance of birdlife throughout the hemisphere, far into the future. This group's report, *Saving Our Shared Birds*, provides a wealth of information accessible to students. This final Investigation is structured around six types of conservation action outlined in *Saving Our Shared Birds* and deemed essential for conserving species and habitats at greatest risk.

Standards Addressed in This Investigation

Next Generation Science Standards – Middle School

- Analyze and interpret data to provide evidence for the effects of resource availability on organisms and populations of organisms in an ecosystem. (MS-LS2-1)
- Construct an explanation that predicts patterns of interactions among organisms across multiple ecosystems. (MS-LS2-2)
- Evaluate competing design solutions for maintaining biodiversity and ecosystem services. (MS-LS2-5)
- Construct an argument supported by evidence for how increases in human population and per-capita consumption of natural resources impact Earth's systems. (MS-ESS3-4)
- Define the criteria and constraints of a design problem with sufficient precision to ensure a successful solution, taking into account relevant scientific principles and potential impacts on people and the natural environment that may limit possible solutions. (MS-ETS1-1)

Next Generation Science Standards – High School

- Design, evaluate, and refine a solution for reducing the impacts of human activities on the environment and biodiversity. (HS-LS2-7)
- Evaluate or refine a technological solution that reduces impacts of human activities on natural systems. (HS-ESS3-4)
- Design a solution to a complex real-world problem by breaking it down into smaller, more manageable

problems that can be solved through engineering. (HS-ETS1-2)

Next Generation Science Standards – Scientific Processes

- Asking questions and defining problems
- Planning and carrying out investigations
- Constructing explanations and designing solutions
- Engaging in argument from evidence
- Obtaining, evaluating, and communicating information

Common Core (9-10th grade)

Reading Science & Technical Subjects

Key Ideas and Details

- Cite specific textual evidence to support analysis of science and technical texts, attending to the precise details of explanations or descriptions. (RST.9-10.1)
- Determine the central ideas or conclusions of a text; trace the text's explanation or depiction of a complex process, phenomenon, or concept; provide an accurate summary of the text. (RST.9-10.2)

Craft and Structure

- Determine the meaning of symbols, key terms, and other domain-specific words and phrases as they are used in a specific scientific or technical context relevant to grades 9-10 texts and topics. (RST.9-10.4)

Range of Reading and Level of Text Complexity

- By the end of grade 10, read and comprehend science/technical texts in the grades 9-10 text complexity band independently and proficiently. (RST.9-10.10)

Writing

Production and Distribution of Writing

- Use technology, including the Internet, to produce, publish, and update individual or shared writing products, taking advantage of technology's capacity to link to other information and to display information flexibly and dynamically. (W.9-10.6)

Research to Build and Present Knowledge

- Conduct short as well as more sustained research projects to answer a question (including a self-generated question) or solve a problem; narrow or broaden the inquiry when appropriate; synthesize multiple sources on the subject, demonstrating understanding of the subject under investigation. (W.9-10.7)
- Gather relevant information from multiple authoritative print and digital sources, using advanced searches effectively; assess the usefulness of each source in answering the research question; integrate information into the text selectively to maintain the flow of ideas, avoiding plagiarism and following a standard format for citation. (W.9-10.8)
- Draw evidence from literary or informational texts to support analysis, reflection, and research. (W.9-10.9)

Speaking and Listening

Comprehension and Collaboration

- Initiate and participate effectively in a range of collaborative discussions (one-on-one, in groups, and teacher-led) with diverse partners on grades 9-10 topics, texts, and issues, building on others' ideas and expressing their own clearly and persuasively. (SL.9-10.1)

- Integrate multiple sources of information presented in diverse media or formats (e.g., visually, quantitatively, orally) evaluating the credibility and accuracy of each source. (SL.9-10.2)

Presentation of Knowledge and Ideas

- Present information, findings, and supporting evidence clearly, concisely, and logically such that listeners can follow the line of reasoning and the organization, development, substance, and style are appropriate to purpose, audience, and task. (SL.9-10.4)

- Make strategic use of digital media (e.g., textual, graphical, audio, visual, and interactive elements) in presentations to enhance understanding of findings, reasoning, and evidence and to add interest. (SL.9-10.5)

Creating a Conservation Plan

Before You Start...

Time

Preparation: **25 minutes**

Instruction: **90-120 minutes**

Preparation

- Download the *Saving Our Shared Birds* report (http://crossingboundaries.org/bwb.php) and make copies of the pages indicated in the Materials list.

- Print copies of the Student Readings, Worksheets, and Presentation Guidelines. Note: students will work in groups on the Primary Actions, and each group will need only the reading and worksheet applicable to their designated Primary Action.

Technology

A-V Equipment

- LCD Projector
- Screen or Whiteboard

Computers

- Acrobat Reader

Information and Communication Technology

- Capzles or other multimedia presentation options

Low-Tech Options

- Have students use PowerPoint or similar presentation software that doesn't require an internet connection.

Materials

Student Readings from Saving Our Shared Birds (the first reading is for all students and the others are for groups)

- A Call to Tri-National Action (pp. 22-35)
- Primary Action 1 (pp. 23-25)
- Primary Action 2 (pp. 26-28)
- Primary Action 3 (p. 29)
- Primary Action 4 (pp. 30-31)
- Primary Action 5 (pp. 32-33)
- Primary Action 6 (pp. 34-35)

Worksheets (for all students)

- Conservation Action
- Presentation Guidelines

Primary Action Worksheets (for groups)

- Protect and Recover Species at Greatest Risk
- Conserve Habitats and Ecosystem Functions
- Reduce Bird Mortality
- Expand Our Knowledge Base for Conservation
- Engage People in Conservation Action
- Increase the Power of International Partnerships

Creating a Conservation Plan
The Investigation

Investigation Overview

1. Explore bird conservation needs and efforts in the U.S. and neighboring countries, and select one of six primary types of conservation action to investigate in depth.

2. Investigate one of six primary actions for conservation and protection of bird life.

3. Design a conservation plan tailored to one of the six focal areas.

4. Create a multimedia presentation to educate the public on this topic and approach.

Learning Objectives

Students will be able to:

- Describe threats to bird species and their habitats.

- Evaluate ways to protect, restore, and enhance populations and habitats of North American birds.

- Create and present an action plan to address critical conservation needs.

Key Concepts

- Biodiversity
- Conservation
- Adaptation
- Critical habitats
- Human impacts
- Threats to biodiversity

Technology Overview

- Students select from a variety of options for creating and presenting a multimedia-rich conservation plan.

Conducting the Investigation

1. Explore bird conservation needs and efforts in the U.S. and neighboring countries, and select one of six primary types of conservation action to investigate in depth.

   a. As a class, view the brief bird conservation video (http:// crossingboundaries.org/bwb.php).

   b. Ask students to use the **Saving Our Shared Birds** report to answer the questions in the **Conservation Action Worksheet.** Through reading designated pages and answering these questions, they will become familiar with the six primary actions identified for tri-national action.

Bird Conservation Report

2. Investigate one of six primary actions for conservation and protection of bird life.

   a. Individually or in small groups, have students select one of the six primary actions as a focus for their investigation and presentation.

      1) Protect and recover species at greatest risk. Students will select from

a list of bird species of high tri-national concern.

    2) <u>Conserve habitats and ecosystem functions</u>. Students will concentrate on sustainable practices needed to conserve bird life in a specific ecosystem.

    3) <u>Reduce bird mortality</u>. Students will explore human-caused threats affecting bird migration, such as pesticides, structural collisions, etc.

    4) <u>Expand our knowledge base for conservation</u>. Students will investigate upcoming trends and topics for scientific research.

    5) <u>Engage people in conservation action</u>. Students will investigate ways in which the U.S., Canada, and Mexico can work together to educate the public in creating and implementing solutions to conservation strategies.

    6) <u>Increase the power of international partnerships</u>. Students will look for ways to expand international partnerships and develop ideas for achieving conservation.

  b. Provide students with the appropriate worksheet (Primary Actions #1-6) to guide investigation of their selected topic.

  c. Have students investigate their primary action, taking notes from relevant sections of the *Saving Our Shared Birds* report and other sources.

  d. Ask students to create an outline for a conservation plan that is tailored to their topic of choice.

3. Create a multimedia presentation to educate the public on the selected topic and conservation approach.

  a. Using the structure provided in the **Presentation Guidelines** handout, ask students to think about how to "sell" their specific conservation plan within the context of these two questions:

- What specific threats are birds facing, in the U.S. and beyond our borders?

- What conservation efforts will protect or restore one or more species or their critical habitats?

  b. Have each student or group create a presentation using a multimedia tool such as Capzles (http://www.capzles.com) to make a convincing case for their conservation plans. (With Capzles, they can create a presentation and upload documents, photos, graphs, videos, sound files, and blog posts. Many alternatives exist, such as Prezi, PowerPoint, or Glogster, but Capzles provides the option to share student presentations with teacher-defined privacy options. This makes it easy to share class-to-class, upload to a wiki, post on Facebook, or share through other social media sites.)

http://www.capzles.com

4. Present conservation plans.

  a. Provide time for student presentations, either projected to the class as a whole or accessed online in a computer lab.

  b. Hold a wrap-up discussion and vote as a class on the highest priority conservation strategy, or delay this wrap-up until after implementing one of the optional extensions below.

5. Possible extensions:

    a. Conduct a conservation project. Ideally this would be based on the actions recommended by students but could also be modified into a more small-scale project feasible within whatever timeframe you have available. Possibilities include habitat improvement in your schoolyard or community, or educating the public or younger students about ways in which they can make a difference in the lives of birds. Your class can become part of a public record of conservation actions by youth by posting on the Lab of Ornithology's BirdSleuth Action Map (http://www.birdsleuth.org/action-map/) to inform and inspire others.

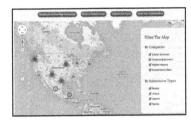

BirdSleuth

    b. Correspond with other students. Have students correspond with students in other classes or even other countries about birds and conservation. Consider using ePals Global Community (http://epals.com) to connect your classes with students in another country. This networking site enables teachers and students to interact with other schools around the world. Teachers can use ePals to find colleagues with similar interests, for example teachers in the U.S. or abroad whose classes are investigating birds and conservation issues. After teachers arrange partnerships, their students exchange emails with a pen pal, and teachers have the option of vetting all incoming and outgoing correspondence.

ePals

CROSSING
boundaries

Name: _____ Period: _____ Date: _____

Conservation Action
Worksheet

The *Saving Our Shared Birds* report lists six types of actions needed to protect, restore, and enhance populations and habitats of North America's birds. Using selected pages of this report, write one sentence answering the following questions about each type of action.

1. <u>Protect and recover species at greatest risk</u>. Many of the species at greatest risk have very limited distributions in highly threatened habitats. Why do these species require individual attention rather than relying on more general conservation actions?

    (Pages 23-25)

2. <u>Conserve habitats and ecosystem functions</u>. Why is this important for bird conservation?

    (Pages 26-28)

3. <u>Reduce bird mortality</u>. What types of challenges do birds face other than loss of habitat?

    (Page 29)

Investigation 8

4. <u>Expand our knowledge base for conservation</u>. What kinds of research are needed to better understand the needs of birds?

(Pages 30-31)

5. <u>Engage people in conservation action</u>. One of the possibilities is to engage more people in citizen science. How can this help with conservation?

(Pages 32-33)

6. <u>Increase the power of international partnerships</u>. What three countries does this report include, and why are partnerships among these countries important for bird conservation?

(Pages 34-35)

CROSSING
boundaries

Name: _____ Period: _____ Date: _____

Conservation Action
Worksheet Answer Key

The **Saving Our Shared Birds** report lists six types of actions need to protect, restore, and enhance populations and habitats of North America's birds. Using selected pages of this report, write one sentence answering the following questions about each type of action.

1. <u>Protect and recover species at greatest risk</u>. Many of the species at greatest risk have very limited distributions in highly threatened habitats. Why do these species require individual attention rather than relying on more general conservation actions?

   (Pages 23-25)

   For species with highly specific habitat requirements, conservation needs to focus on protecting specific sites that provide critical habitat.

2. <u>Conserve habitats and ecosystem functions</u>. Why is this important for bird conservation?

   (Pages 26-28)

   Habitat conservation is needed to keep common species common and to reverse population declines in species that are at risk.

3. <u>Reduce bird mortality</u>. What types of challenges do birds face other than loss of habitat?

   (Page 29)

   Birds are caught and sold as pets, they are killed by cats and by collisions with windows or other structures, and they die from exposure to pesticides.

4. <u>Expand our knowledge base for conservation</u>. What kinds of research are needed to better understand the needs of birds?

(Pages 30-31)

Research is needed on population status and trends, habitats, limiting factors, and causes of population declines for bird species of high tri-national concern.

5. <u>Engage people in conservation action</u>. One of the possibilities is to engage more people in citizen science. How can this help with conservation?

(Pages 32-33)

Citizen science provides valuable scientific data and also helps people to learn about birds and can motivate them to be involved in conservation.

6. <u>Increase the power of international partnerships</u>. What three countries does this report include, and why are partnerships among these countries important for bird conservation?

(Pages 34-35)

The report includes Canada, the U.S. and Mexico, and these three countries share many bird species, habitats, and threats.

Name: _____ Period: _____ Date: _____

Primary Action #1 Protect and Recover Species at Greatest Risk (pages 23-25)

1. List the habitats of landbird species.

2. Where is a strong network of protected areas needed?

3. What is the common link between the Wood Thrush and the Slaty-tailed Trogon? Why is this link important?

4. What are some ways that species at risk can make recoveries?

5. List some of the agencies that coordinate cross-country conservation efforts.

6. Choose one of these bird species to focus on for your conservation plan, and then investigate ways to protect or recover this species.

__ Spotted Owl	__ Sprague's Pipit	__ Baird's Sparrow
__ Chestnut-collared Longspur	__ Rufous Hummingbird	__ Loggerhead Shrike
__ Brewer's Sparrow	__ Field Sparrow	__ Lark Bunting
__ Harris's Sparrow	__ Cassin's Finch	

Primary Action #1 Protect and Recover Species at Greatest Risk (pages 23-25) Answer Key

1. List the habitats of landbird species.

    - *Tropical highland and pine-oak forest of the Mexican mountains*
    - *Tropical deciduous forests on the Pacific slope of Mexico*
    - *Tropical evergreen forest from southern Mexico through Central America*

2. Where is a strong network of protected areas needed?

    Along Mexico's Pacific Coast and in narrow highland regions from Tamaulipas and Chihuahua south to Chiapas

3. What is the common link between the Wood Thrush and the Slaty-tailed Trogon? Why is this link important?

    The Wood Thrush spends winter in tropical evergreen forests of southern Mexico, the same forests where the Slaty-tailed Trogon lives year-round.

4. What are some ways that species at risk can make recoveries?

    - *Protecting sufficient amounts of critical habitat for endangered species throughout their life cycles.*
    - *Implementing the recovery components of endangered species laws and other wildlife conservation legislation.*

5. List some of the agencies that coordinate cross-country conservation efforts.

    - *Mesoamerican Pine-Oak Conservation Alliance*
    - *Pronatura Sur*
    - *The Nature Conservancy*

6. Choose one of these bird species to focus on for your conservation plan, and then investigate ways to protect or recover this species.

__ Spotted Owl	__ Sprague's Pipit	__ Baird's Sparrow
__ Chestnut-collared Longspur	__ Rufous Hummingbird	__ Loggerhead Shrike
__ Brewer's Sparrow	__ Field Sparrow	__ Lark Bunting
__ Harris's Sparrow	__ Cassin's Finch	

Name: _____ Period: _____ Date: _____

Primary Action #2 Conserve Habitats and Ecosystem Functions (pages 26-28)

1. Explain how the expansion of agriculture is threating biodiversity.

2. According to this article, what should "restoration" promote?

3. Give an example of sustainable grazing and food production that will benefit bird species.

4. List three ecosystems that are being degraded at an alarming rate.

   a)

   b)

   c)

5. Identify one sustainable forestry practice.

6. Describe a sustainable community-based initiative.

7. Choose one of the following habitats to focus on in your conservation plan:

   __Tropical Evergreen Forest   __Boreal Forest   __Temperate Grassland
   __Arid Lands   __Temperate Forest   __Urban

Name: _____ Period: _____ Date: _____

Primary Action #2 Conserve Habitats and Ecosystem Functions (pages 26-28) Answer Key

1. Explain how the expansion of agriculture is threating biodiversity.

   Expansion of areas used for livestock grazing and other forms of agriculture affects the grassland, forest, and arid habitats needed by many species.

2. According to this article, what should "restoration" promote?

   - *Use of native plant species*
   - *Control of invasive species*
   - *Minimizing the use of chemicals*
   - *Using fire to emulate natural disturbance patterns*

3. Give an example of sustainable grazing and food production that will benefit bird species.

   Shade-grown coffee and cocoa helps to retain native forest cover for birds.

4. List three ecosystems that are being degraded at an alarming rate.

   a) *Native grasslands*
   b) *Temperate broadleaf forests*
   c) *Tropical dry forests*

5. Identify one sustainable forestry practice.

   Possibilities include:

   - *Retaining large trees and snags*
   - *Using fire and other disturbances to create greater forest complexity*
   - *Managing to maintain structural complexity and diversity of age classes*

6. Describe a sustainable community-based initiative.

   Resource management is integrated into the regional economy by working with local communities and landowners to protect, restore, and manage habitats. This could include taxation systems, market incentives, and conservation easements.

7. Choose one of the following habitats to focus on in your conservation plan:

__Tropical Evergreen Forest	__Boreal Forest	__Temperate Grassland
__Arid Lands	__Temperate Forest	__Urban

CROSSING
boundaries

Name: _____ Period: _____ Date: _____

Primary Action #3 Reduce Bird Mortality
(page 29)

1. Explain cage bird trade.

2. How can this trade be regulated?

3. What are some types of human-made structures that impact migratory birds?

4. What two websites provide information on minimizing bird collisions?

5. How can we minimize deaths from pesticide use?

6. Select one source of human-caused bird mortality to investigate for your conservation plan. What type of bird mortality have you selected, and why do you think it is important?

Name: _____ Period: _____ Date: _____

Primary Action #3 Reduce Bird Mortality
(page 29) Answer Key

1. Explain cage bird trade.

   Many species of birds are trapped and sold. These include all Mexican parrot species, three species of toucans, many orioles, and buntings.

2. How can this trade be regulated?

   Trapping quotas could be set, tied to effective monitoring of wild bird populations. This would be similar to the successful model of sustainable waterfowl harvesting that has been achieved under the North American Waterfowl Management Plan.

3. What are some types of human-made structures that impact migratory birds?

   Windows, wind turbines, transmission lines, and other infrastructure

4. What two websites provide information on minimizing bird collisions?

   - *www.flap.org for minimizing window kills*
   - *www.aplic.org for bird-friendly lighting of towers and buildings, and protecting birds on power lines*

5. How can we minimize deaths from pesticide use?

   Supporting organic agriculture, developing reduced use or lower toxicity alternatives, and developing shared standards for licensed pesticides and application techniques

6. Select one source of human-caused bird mortality to investigate for your conservation plan. What type of bird mortality have you selected, and why do you think it is important?

Name: _____ Period: _____ Date: _____

Primary Action #4 Expand Our Knowledge Base for Conservation (pages 30-31)

1. What are some basic habitat and ecological requirements that must be studied for effective bird conservation?

2. What technologies are available to provide data on population status and trends? What do they allow us to do?

3. What makes it difficult to know the population status of many species?

4. What are some examples of human causes of bird mortality, and why is it important to study these sources?

5. What is meant by the "human dimensions of bird conservation," and why is this an important topic for research?

6. Select a research topic to focus on for your conservation plan. This should be a topic for which better understanding will be useful in designing and implementing effective conservation efforts. What is your topic, and why is it important?

Primary Action #4 Expand Our Knowledge Base for Conservation (pages 30-31) Answer Key

1. What are some basic habitat and ecological requirements that must be studied for effective bird conservation?

    Information is needed about the food, vegetation, and patch size critical for priority species in all habitats, especially in rapidly diminishing tropical forest.

2. What technologies are available to provide data on population status and trends? What do they allow us to do?

    Devices such as geolocators are useful in tracking migration routes. Study of stable isotopes in feathers is another technique being used to document migration patterns.

3. What makes it difficult to know the population status of many species?

    Monitoring has been limited in areas where access is difficult, remote, or expensive.

4. What are some examples of human causes of bird mortality, and why is it important to study these sources?

    Examples include collisions with tall structures and vehicles, and predation by cats and other nonnative predators. Studying these sources is important because we need to understand the effects on bird populations in order to set policies and guidelines based on the relative risks to priority species.

5. What is meant by the "human dimensions of bird conservation," and why is this an important topic for research?

    "Human dimensions" refers to how and why people relate to birds and bird conservation issues. We need to understand peoples attitudes and behaviors in order to design conservation solutions that are acceptable to society.

6. Select a research topic to focus on for your conservation plan. This should be a topic for which better understanding will be useful in designing and implementing effective conservation efforts. What is your topic, and why is it important?

Name: _____ **Period:** _____ **Date:** _____

Primary Action #5 Engage People in Conservation Action (pages 32-33)

1. Why is it important to engage people in conservation action?

2. Who are the stakeholders that must be reached to implement bird friendly agriculture and forestry practices?

3. What citizen science projects are mentioned here as effective ways to engage volunteers in bird monitoring?

4. Give an example of a bird-friendly economic opportunity that could be used in a rural community, and say why this is important.

5. What does coffee farming have to do with bird habitat?

6. Choose one of the following websites from the article. In the conservation plan that you create, include information on how this group is helping to engage and educate people about birds and conservation:

   ___ Environment for the Americas (www.birdday.org)

   ___ eBird (www.ebird.org)

   ___ Colorado Birding Trail (www.coloradobirdingtrail.com)

   ___ El Fondo de Conservación El Triunfo A.C. (www.fondoeltriunfo.org)

Investigation 8

Primary Action #5 Engage People in Conservation Action (pages 32-33) Answer Key

1. Why is it important to engage people in conservation action?

   We need to motivate people to recognize the costs and benefits of alternative futures, make behavioral changes, and take conservation actions. Through our actions, humans continue to threaten birds, often unknowingly.

2. Who are the stakeholders that must be reached to implement bird friendly agriculture and forestry practices?

   Producers, industry, policy-makers, business communities, and first Nations, Native American, and indigenous peoples

3. What citizen science projects are mentioned here as effective ways to engage volunteers in bird monitoring?

   The North American Breeding Bird Survey, Christmas Bird Count, and eBird

4. Give an example of a bird-friendly economic opportunity that could be used in a rural community, and say why this is important.

   Examples of bird-friendly economic opportunities include ecotourism and birding festivals. Providing bird-friendly economic opportunities reduces threats in and around protected areas.

5. What does coffee farming have to do with bird habitat?

   Coffee is an important crop in the tropics. On modern intensive coffee farms, tropical forests have been replaced with fields containing just coffee plants. The traditional method of growing coffee in the shade protects tropical forests and provides important habitat for many resident and migratory bird species.

6. Choose one of the following websites from the article. In the conservation plan that you create, include information on how this group is helping to engage and educate people about birds and conservation:

   ___ Environment for the Americas (www.birdday.org)

   ___ eBird (www.ebird.org)

   ___ Colorado Birding Trail (www.coloradobirdingtrail.com)

   ___ El Fondo de Conservación El Triunfo A.C. (www.fondoeltriunfo.org)

Name: _____ Period: _____ Date: _____

Primary Action #6 Increase the Power of International Partnerships (pages 34-35)

1. What does the NABCI promote?

2. List several areas in which International joint ventures hope to increase support and expand their capacity.

3. What will help government-led programs succeed?

4. Give an example of a voluntary partnership for high concern species and briefly discuss their role.

5. Choose an international organization that is currently trying to increase tri-national efforts in bird conservation. In the conservation plan that you create, discuss the methods used by this groups to involve partnering countries. Here are a few organizations mentioned in the article, but feel free to include another of your choice:

    __ el Grupo Cerúleo (www.srs.fs.usda.gov/egc)

    __ North American Bird Conservation Initiative (www.nabci.net)

    __ Avian Knowledge Network (www.avianknowledge.net)

Primary Action #6 Increase the Power of International Partnerships (pages 34-35) Answer Key

1. What does the NABCI promote?

   The North American Bird Conservation Initiative promotes a strategic approach to conserving birds through the identification of continentally important areas.

2. List several areas in which International joint ventures hope to increase support and expand their capacity.

   In the aridlands, tropical deciduous forests, and pine-oak forests of the southwestern U.S. and northern Mexico.

3. What will help government-led programs succeed?

   Increased funding for existing programs, including from private foundations, international aid organizations, and industry partner

4. Give an example of a voluntary partnership for high concern species and briefly discuss their role.

   The Cerulean Warbler Technical Group brings together partners from the forest-products, coal-mining, and coffee-production industries, multiple resource agencies, conservation NGOs (non-government organizations), and university scientists from countries throughout the species' breeding and winter range. Together they are working to protect this vulnerable species.

5. Choose an international organization that is currently trying to increase tri-national efforts in bird conservation. In the conservation plan that you create, discuss the methods used by this groups to involve partnering countries. Here are a few organizations mentioned in the article, but feel free to include another of your choice:

   ___ el Grupo Cerúleo (www.srs.fs.usda.gov/egc)

   ___ North American Bird Conservation Initiative (www.nabci.net)

   ___ Avian Knowledge Network (www.avianknowledge.net)

Name: _____ Period: _____ Date: _____

CROSSING
boundaries

Tri-National Bird Conservation Plan Presentation Guidelines

As you create your presentation, think about how to convince people that your conservation plan is important. Use this structure and check off items as they're completed.

Include the following information:

☐ **Slide 1:** Include your name, the title of the Primary Action you are addressing, and a photo or other visual if you'd like.

☐ **Slides 2 & 3:** What specific threats do birds face, in the U.S. and beyond our borders?

☐ **Slide 4:** Information from your worksheet. Include important facts or figures from your research and readings.

☐ **Slides 5 & 6:** What conservation efforts do you propose in order to protect or restore one or more bird species or critical habitats?

☐ **Slides 7 & 8:** Maps, charts, pictures that support your argument or position.

☐ **Slide 9:** Citations and acknowledgments.

Investigation 1

Taza Schaming. Clark's Nutcracker. Cornell Lab of Ornithology.

Kelly Colgan Azar. Downy Woodpecker. (https://flic.kr/p/atLcYx)

Kelly Colgan Azar. Magnolia Warbler. (https://flic.kr/p/aimqEE)

Investigation 2

Kelly Colgan Azar. Belted Kingfisher. (https://flic.kr/p/8oAmcS)

Kelly Colgan Azar. Wood Thrush. (https://flic.kr/p/8aFWp6)

Kelly Colgan Azar. Red-Bellied Woodpecker. (https://flic.kr/p/dN8pDp)

Investigation 3

Kevin J. McGowan. Northern Parula. Cornell Lab of Ornithology.

Phil Kahler. Northern Pintail. (https://flic.kr/p/dNVHZE)

Andy Johnson. Hudsonian Godwit. (andyjohnsonphoto.com)

Investigation 4

Kevin J. McGowan. Painted Bunting. Cornell Lab of Ornithology.

Kevin J. McGowan. Northern Cardinal. Cornell Lab of Ornithology.

Kevin J. McGowan. Scarlet Tanager. Cornell Lab of Ornithology.

Investigation 5

Phil Kahler. Bald Eagle. (https://flic.kr/p/mFyEaH)

Phil Kahler. Lazuli Bunting. (https://flic.kr/p/nx9w6w)

Kevin J. McGowan. Indigo Bunting. Cornell Lab of Ornithology.

Investigation 6

Phil Kahler. Silver Beaked Tanager. (https://flic.kr/p/o5zDRL)

Phil Kahler. Atlantic Puffin. (https://flic.kr/p/ov2FBt)

Phil Kahler. Green Honeycreeper. (https://flic.kr/p/oiE9k5)

Investigation 7

Phil Kahler. Snowy Owl. (https://flic.kr/p/dNPF1B)

Phil Kahler. Long-Billed Dowitcher. (https://flic.kr/p/dRf59p)

Phil Kahler. Black Turnstone. (https://flic.kr/p/oXzCzk)

Investigation 8

Phil Kahler. Spotted Owl. Personal Collection.

Phil Kahler. Pileated Woodpecker. (https://flic.kr/p/nDtr1Z)

Phil Kahler. Western Meadowlark. (https://flic.kr/p/e87Jwz)

Permitted Uses, Prohibited Uses and Liability Limitation

Permitted Uses of Materials

The materials from this book and associated online materials are provided for educational purposes only, for the convenience of the purchasing teacher only for use in his or her lesson planning, teaching, or related educational activities. Other digital materials provided online (such as files or videos) are intended to be used by teachers or students to compliment the lessons provided in this book. You may use, copy, reproduce, and distribute the materials only in quantities sufficient to meet the reasonable needs of your classroom and students.

Prohibited Uses of Digital Materials

You may not sell, rent, lease, sublicense, loan, assign, time-share, or transfer, in whole or in part, any portion of this material except as stated above. You may not remove or obscure any copyright or trademark notices of Carte Diem Press as an entity of Critical Think Inc. You may not enter into any transfer or exchange of material except as provided for in herein. You may not use the digital materials provided, whether in digital or tangible form, except in conjunction with the exercises and context of this book.

You may not create any derivative works from the digital materials, except for your own noncommercial use in your classroom in conjunction with the exercises and context of this book, as provided herein.

Limitation of Critical Think Inc.'s Liability

As outlined in the metadata for each map document and layer package stored on ArcGIS Online, Carte Diem Press as an entity of Critical Think Inc., shall not be liable for direct, indirect, special, incidental, or consequential damages related to use of the digital materials, even if Carte Diem Press as an entity of Critical Think Inc., is advised of the possibility of such damage. Any data used is posted specifically for use with Birds without Borders and is meant for educational purposes only.

ABOUT the AUTHORS

Dr. Nancy Trautmann

Dr. Nancy Trautmann is Director of Education at the Cornell Lab of Ornithology, where she leads a team that aims to inspire curiosity, learning, and action related to birds and the natural world.

Dr. Jim MaKinster

Dr. Jim MaKinster is a Professor of Science Education at Hobart and William Smith Colleges, where he teaches a variety of courses in science and environmental education and leads projects focused on environmental issues, scientific inquiry, and the use of geospatial technology.

Together, they lead programs for middle and high school teachers. The investigations in this book were developed in collaboration with 60 teachers who participated in the NSF-funded Crossing Boundaries project. These teachers pioneered new uses of technology in their classrooms and provided key input about successes, challenges, and adaptations for various types of courses and students.

Crossing Boundaries is a collaborative effort between Hobart and William Smith Colleges and the Cornell Lab of Ornithology. With curriculum, data resources, and related professional development opportunities, Crossing Boundaries equips teache to engage students in using information and communication technologies (ICT) to address biodiversity conservation issues from local to international in scope.

CARTE DIEM PRESS
map the day

ALWAYS UP-TO-DATE

You never have to worry about the lessons "not working," because we're working hard to keep it current. **We're committed to sustainable curriculum.**

We know how fast technology changes and how important reliable curriculum is. Register your book and we will contact you when we have updated content. We'll send you a complimentary digital version of the updates.

Stay in the loop!
Register your copy today
http://gisetc.com/register